Shallot and Nitrogen Fertilizer

Tiru Tesfa

London·Istanbul·Moscow·Delhi·Jakarta

Shallot and Nitrogen Fertilizer
by Tiru Tesfa

Printed in the UK

Published by Glimmer Publishing Ltd. Ground Floor 2, Woodberry Grove, London, N12 0DR, England.

Glimmer books may be purchased for educational, business, or sales promotional use. Online editions are also available for most titles (*http://glimmerpublishing.com*). For more information, contact our corporate sales department: *sales@glimmerpublishing.com*.

April 2018: First Edition

See http://glimmerpublishing.com/978-1-78902-000-7 for release details

The Glimmer logo is a registered trademark of Glimmer Publishing Ltd.
The cover image by www.gardenandgardener.co.uk
Cover design by Glimmer Publishing Ltd.

ISBN: 978-1-78902-000-7

SHALLOT
AND
NITROGEN FERTILIZER

TIRU TESFA
University of Gondar
Gondar

DEDICATION

This book is dedicated to my parents:

Mr. Tesfa Belachew, who did not see the fruit of his seedling

and

Ms. Abetir Yismaw, for guiding me towards success in my life.

BIOGRAPHICAL SKETCH

Tiru Tesfa was born on April 11, 1980 in Bahir Dar, the capital city of Amhara National Regional State, Ethiopia. He attended elementary school at Kebele 03 primary school and Shinbit primary school from 1986-91 and secondary school at Fasilo compulsory secondary school and Tana Haik secondary school from 1992-97 in Bahir Dar town. He joined Haramaya University, the then Alemaya University, and graduated in 2002 with a B.Sc. degree in Plant Sciences and in 2008 with a M.Sc. degree in Horticulture. After graduation he was employed by Amhara Regional Agricultural Research Institute, Sirinka Agricultural Research Center and served as an assistant researcher from 2003-2010. The author is working in University of Gondar, Department of Horticulture since March 2014.

ACKNOWLEDGEMENTS

First and foremost, I wish to extend my sincere and heartfelt appreciation and deepest gratitude to Prof. Kebede W/Tsadik and Dr Wondimu Bayu for their professional guidance bestowed in reading and commenting of draft book, elaborating their valuable comments and constructive criticism.

Above all, Loving-Kindness and faithfulness of almighty God and His Mother St. Merry is bestowing health, strength, patience and protection throughout the preparation of this book is highly appreciated. Let God bless the coming days, Amen!!!

LIST OF ACRONYMS AND ABBREVIATIONS

ARARI	Amhara Regional Agricultural Research Institute
CACC	Central Agricultural Census Commission
CEC	Cation Exchange Capacity
DM	Dry Matter
DMRT	Duncan's Multiple Range Test
G	Genotype
LSD	Least Significance Difference
OC	Organic carbon
OM	Organic Matter
SARC	Sirinka Agricultural Research Center
TSS	Total Soluble Solids

TABLE OF CONTENTS

LIST OF TABLES

LIST OF TABLES IN THE APPENDIX

LIST OF FIGURES IN THE APPENDIX

NITROGEN FERTILIZER ON YIELD, YIELD COMPONENTS AND SHELF LIFE OF SHALLOT

ABSTRACT

Field and storage studies on effects of nitrogen fertilization on shallot (Allium cepa var ascalonicum Baker) genotypes were undertaken in Wello, Northeastern Ethiopia during 2007/08. The field trial was done on silt-sandy soil with very low organic matter content under irrigation. The storage trial was undertaken in a grass-thatched hut under ambient condition. The objectives of this study were to investigate the effect of nitrogen fertilizer rates on yield and yield components, and to determine the level of nitrogen fertilization at which shallot bulbs could be stored for optimum periods of time. The treatments comprised factorial combinations of four levels of nitrogen (0, 50, 100 and 150 kg N ha^{-1}) and four genotypes (Huruta, Negelle, Dz-sht-68 and Local) arranged in a randomized complete block design with three replications. Nitrogen fertilization highly significantly (p<0.01) increased all of the growth parameters considered in this study. Similarly, very highly significant (p<0.001) genotypic variations were observed for growth parameters viz: plant height, number of leaves per plant, leaf diameter, biomass per plant and days to maturity. Nitrogen fertilization increased the marketable and total bulb yield without significantly affecting the storability of bulbs. Number of bulb splits per plant, bulb diameter, mean bulb weight, unmarketable bulb yield and percent bulb weight loss were also significantly (p<0.05) influenced by interaction effects of nitrogen fertilization and genotypes. The interaction effect of N at 150 kg ha^{-1} increased the mean bulb weight of the local cultivar by 175.2% and 108.8% compared to local cultivar fertilized with 0 and 50 kg N ha^{-1}, respectively. Nitrogen fertilization at 150 kg N ha^{-1} increased the marketable and total bulb yields in comparison with unfertilized control by 25.77% and 25.82%, respectively. The highest marketable (28.31 t ha^{-1}) and total (29.31 t ha^{-1}) bulb yields were obtained in a plot fertilized with 150 Kg N ha^{-1} in comparison with unfertilized control, but the 100 kg N ha^{-1} gave marketable (25.84 t ha^{-1}) and total (26.86 t ha^{-1}) bulb yields that were statistically similar to the highest yields. Genotypic bulb yield variations were also observed where the highest and the lowest marketable bulb yields were obtained in Dz-sht-68 (31.34 t ha^{-1}) and local (13.22 t ha^{-1}) genotypes, respectively. The result of this study revealed that N rate of 100 kg ha^{-1} was found to be optimum for shallot bulb production without significantly affecting quality (Pungency, TSS and DM) and short-term storability of the bulbs. On the contrary, the shallot genotypes studied showed variation in fresh bulb yield and quality (TSS and DM) without significant difference in storability of the bulbs. In this study application of N fertilizer did not affect the shelf life of shallot genotypes used, but overall storability of the bulbs was very low due to high temperature and low RH suggesting the need for further investigation to improve the storage problem of shallot bulbs.

Key words: shallot, nitrogen fertilizer, genotype, yield, yield components, shelf life.

1. INTRODUCTION

Shallot (*Allium cepa* var *ascalonicum* Baker; syn. *A. ascalonicum auct.* Non L.) belongs to the genus Allium and family Alliaceae. The genus Allium is distributed from the tropics to subarctic belt, but a region of high species diversity stretches from the Mediterranean basin to Central Asia and Pakistan (Fritsch and Friesen, 2002). As shallot and its relative species are generally open pollinated crops and have been cultivated for long time, a number of land races and natural hybrids either intraspecific or interspecific probably are to be on the increase (Arifin and Okubo, 1996). The majority of shallot genotypes are clonally propagated, even where seed production is possible, to maintain the unique quality traits and population homogeneity of highly heterogeneous plant (Currah and Proctor, 1990).

Shallots are important alliaceous crops cultivated in many tropical countries as a substitute for bulb onions (*Allium cepa* L. var *cepa*). Although bulb onions can be grown in the tropics, farmers in the tropical countries prefer shallots to the common onions for their shorter growth cycle, better tolerance to disease and drought stresses, longer storage life and for their distinct flavour that persists after cooking (Brewster, 1990; Currah and Proctor, 1990; Grubben, 1994; Pathak, 1994; Sumiati, 1994; Abbey *et al.*, 1998).

Shallot is distinguished from bulb onion by its habit of lateral bud growth, however, it appears to be a type of bulb onion, which has been selected for its ability to multiply vegetatively. It was described as a perennial plant that seldom produces seeds. But when the bulb is planted, it divides into a number of bulblets, which remains attached at the bottom. The plant is similar to common onion but bulbs are much smaller in size and may produce from two up to 15 or more bulblets (Thompson and Kelly, 1959; Jones and Mann, 1963).

In Ethiopia, the Alliums group (onion, garlic, and shallot) are important bulb crops produced by small and commercial growers for both local use and export (Yohannes, 1987; Metasebia and Shimels, 1998). These crops are produced for home consumption and as a source of income to many peasant farmers in many parts of the country (Metasebia and Shimels, 1998; Getachew and Asfaw, 2000). Metasebia and Shimels (1998) reported that per capita consumption of these crops is estimated to be over 1.74 kg and 5.9 kg in the rural and urban centre, respectively. Statistics on the production of Allium crops show that about 15,290 ha of land was cultivated and 0.21 million tons of bulbs were produced in the year 2001/2002. The production is spread throughout the country both under irrigation and rainfed conditions in different agro-climatic regions (CACC, 2002).

As the survey report of Getachew and Asfaw (2000) indicated, planting materials used by farmers are usually heterogeneous with respect to size, shape, color, pungency, storability and resistance to diseases. They also differ in their time to reach maturity offering the opportunity to select for different harvest periods and storability. The vegetatively propagated shallot has a very short growing period of only three to four months, which allows it to be grown during the short rains in the dry season. However, productivity of shallot is generally low (about 10-14 t ha^{-1}) (CACC, 2002). This low level of bulb yield compared to experimental plot yields of more than 20 t ha^{-1} could be due to several factors among which moisture stress, low soil fertility, diseases and insect pests, and genotypes are among the lists of constraints for production of shallot in the country (Getachew and Asfaw, 2000).

Nitrogen (N) deficit is the most severe and widespread nutrient constraint limiting the productivity of different crops in Northeastern Ethiopia (Bayu *et al.*, 2002). Mamo *et al.* (1988) also reported that due to a long cropping history and low manure and fertilizer inputs, the nutrient status of Ethiopian soils is generally low and N is the most limiting nutrient for crop production. It is well known that many physiological processes associated with crop growth are enhanced by N supply (Muchow, 1994). Nitrogen plays a central role in plant biochemistry as an essential constituent of cytoplasmic proteins, nucleic acids, chlorophyll, cell walls and a vast array of other cell components (Hay and Walker, 1989). Marschner (1995) described Nitrogen as an important component of proteins, enzymes and vitamins in plants and it is a central part of essential photosynthetic molecule, chloropyll. Consequently, a deficiency in the supply of nitrogen has a profound influence upon crop growth and can lead to a total loss of yield in extreme cases (Hay and Walker, 1989).

Use of plant nutrients are known to affect productivity, quality and storability of onion and shallot cultivars. Nitrogen is the principal plant nutrient required in much greater quantities. Plants demand for N can be satisfied from a combination of soil and fertilizer N to ensure optimum growth. While exogenous N application is known to increase yield of onions and many researchers found that high levels of nitrogenous fertilizer resulted in reduced onion storage life (Bhalekar *et al.*, 1987; Kato *et al.*, 1987; Batal *et al.*, 1994). Shallot is considered to have similar nutritional requirements and its storage life could be affected like other Alliums (Currah and Proctor, 1990; Brewster, 1994; Zaharah *et al.*, 1994; Hussien, 1996; Kebede *et al.*, 2002a, 2003a; Sebsebe, 2006).

Information on nitrogen fertilization and its effect on yield and shelf life is not available for the promising and the recommended shallot cultivars in Ethiopia, specifically for the Eastern part of the Amhara Region. Therefore, this study was initiated with the following objectives:

- to study the effect of nitrogen rate on yield and yield components of some shallot genotypes, and
- to determine the level of nitrogen fertilization at which shallot bulbs could be stored for optimum periods of time.

2. LITERATURE REVIEW

2.1. The Shallot Crop

Shallot is an annual crop with several bulbs arising from a single parent bulb. The bulbs, sometimes referred to as 'cloves', vary in shape, size, and color and consist of a series of overlapping storage leaves arising from condensed conical stem. The foliage leaves are up to 40 cm in length and are slightly flattened on the upper surface. The flower stalks are up to 25 cm and black seeds are produced within a capsule (Jones and Mann, 1963; Rice *et al.*, 1993; Brewster, 1994).

On global scale, shallot is a minor Alliaceous crop; however, in South-East Asia like Indonesia, Sri Lanka, and Thailand as well as in some African countries such as Uganda, Ethiopia and Cote d' Ivoire, where onion seed is hard to produce and its culture is difficult, the vegetatively propagated shallot is cultivated as an important substitute for bulb onion (Currah and Proctor, 1990; Currah, 2002). Shallot is grown dominantly for its dry bulbs, but the tender leaves are also used as salad. It has high dry matter content and strong pungent flavor that is preferred for certain dish preparations such as sauces and soups (Thompson and Kelly, 1959; Currah and Proctor, 1990).

In European countries like France and in the United States of America, shallots are favored for their distinct flavor that persists even after cooking. Shallots have a more delicate flavor than the common onion (Currah and Proctor, 1990). In Ethiopia, shallots are used for flavoring of the local stew 'wot'; they are preferred to bulb onions for their strong pungent flavor in most Ethiopian cuisine. Moreover, they are among the widely cultivated vegetable crops as a source of income by peasant farmers in many parts of the country particularly in the mid and high altitude areas (Getachew, 1996).

The major production areas of shallot in the country are Fedis in Harerge; Huruta, Sire, Sirka, Bekoji and Arsi Negelle in Arsi; Ambo, Wolliso, Godino, Kessem and Majete in Shewa; Bure and the vicinities of Debre Markos in Gojjam; and Waraillu and WaraBabo in Wello (Getachew and Asfaw, 2000).

4

Shallots have a wide range of climatic and soil adaptation and are cultivated both under rainfed and irrigated conditions (Getachew, 1996). Ethiopia, with its wide range of climatic conditions, has the potential to produce a very diverse types of vegetable crops including shallots. The country is considered as a center of diversity for shallot while onion is a recently introduced crop (Seifu, 1981). Shallots are adapted to many soils, loose sandy soils with high organic content are preferable, although silt-clay loams are often used (Rice et al., 1993).

The bulbs tolerate high temperatures up to 30 °C and relatively high temperatures encourage bulb development in most cultivars. With some cultivars, bulbs are not well formed at temperatures lower than 20 °C (Tindall, 1983). Generally, an average optimal temperature of 25-32 °C is required for shallot bulb production during the growing period (Salunkhe and Kadam, 1998).

Altitudes from sea level to 2500 m are considered to be suitable. Although the shallot requires long days for maximum bulb development, most tropical cultivars will form bulbs of an adequate size in short day lengths. Flowering is reduced in high temperature conditions. Yields may be reduced during heavy rainfall, partly due to diseases. A dry period is required for harvesting and curing of the bulbs (Rice et al., 1993).

2.2. Variability in Onion and Shallot Cultivars

Variation is the occurrence of differences among individuals due to the differences in their genetic composition and/or the effects of the environment in which they were raised (Allard, 1960). Altitude, particularly through the influence of climate, strongly affects the growth and yield of shallot. Moreover, soil types strongly affect the bulb quality of onions and shallots in terms of shape, firmness, skin color and dry matter content (Sumiati, 1994).

2.2.1. Phenotypic and genotypic variation

Understanding the extent of genetic and phenotypic variabilities that exist in a crop species is of utmost importance in efforts towards initiating a future breeding program and developing better varieties with full agronomic package. Genetic variability is of immense importance to the breeders and agronomists because it could be transmitted to the progeny and the proper management of this diversity could produce permanent gain in the performance of the plant (Welsh, 1981). In crops like onion, breeders and agronomists focused their efforts on traits important to the production of high quality bulbs, and production of commercially acceptable bulbs depending upon traits such as bulb size, shape, color of skin and flesh, disease resistance, pungency and storability (Havey, 1993).

Ethiopian shallot germplasm collections were reported to vary in shape, color, pungency, storability and other characters (Getachew and Asfaw, 2000). They also indicated that Ethiopian farmers grew different cultivars and named after the major production belt. Mixtures of bolters and non-bolters, and spreading and compact types were observed within a farm. Similarly, Shimels (1998) reported that the popular clones grown in Ethiopia are named and known by the regions or areas in which they are vastly grown, such as Gojjam, Gojjam Bure, Gelemso, Fedis, etc. These names are directly attached to the clonal places of previous diversity. As shallots are propagated mainly by asexual method, it is unclear whether these variabilities reflect a diverse genetic background or not. However, in her recent work with 48 shallot genotypes, Fasika (2004) demonstrated the existence of genetic variability among local collections of shallot genotypes for many important characters including yield and bulb quality.

Barta *et al.* (1983) reported that the attributes showing estimates of variability were earliness, bulb size, dry matter content and bulb weight in onion, and they suggested that high coefficient of genetic variance for yield components could be used to improve cultivars for almost all the characteristics studied. Phenotypic coefficient of variation and genotypic coefficient of variation were observed for leaves per plant, neck thickness, bulb diameter, bulb length, fresh weight above ground, dry weight above ground, yield per plant, harvest index per plant and biological yield per plant in onion (Abayneh, 2001).

Maximum genetic variation was also observed in onion for plant height, leaf length, leaf diameter, neck thickness, fresh weight above ground, dry weight above ground, yield per plant, days to maturity and biological yield per plant. Moderate phenotypic and genotypic coefficients of variation were recorded for number of leaves per plant, bulb length, bulb diameter, bulb dry weight and total soluble solids and the lowest value was observed for days to maturity (Abayneh, 2001). Singh (1981) also reported moderate to high phenotypic and genotypic coefficients of variation for number of leaves per plant, bulb length, bulb diameter, bulb dry weight and total soluble solids in onion.

Varietal differences in nutrient requirements and ability to accumulate and utilize nutrients have been reported for several crops (Epstein and Jefferies, 1964). High yielding varieties usually require more nitrogen than low yielding. These differences could be due to root physiology and development, and the plant ability to mobilize and utilize a given nutrient. On the contrary, Kebede *et al.* (2003a) reported absence of clear effect of N fertilization on yield, quality and storability between improved and local shallot cultivars. Hence, further investigations would be required using available cultivars to generate comprehensive result.

2.2.2. Flavors in onions and shallots

Allium cultivars differ in their flavor intensity. While consumers in many cultures prefer pungent cultivars, others desire cultivars that are mild and sweet (Currah and Proctor, 1990). Similarly, Ketter and Randle (1998) demonstrated that the concentration of flavor precursors in onions are determined by genetics of the cultivar. However, the growing environment can greatly influence flavor intensity of a given cultivar. Shallot is preferred to bulb onions by consumers for its good culinary quality such as high pungency (Grubben, 1994). Alliums are primarily consumed because of their flavors or their ability to enhance the flavor of other foods. Fasika (2004) reported that flavor intensity is a heritable trait in shallots. Lancaster and Boland (1990) observed that onion pungency is influenced by genetic and several environmental factors such as water supply, temperature during growth, sulfate availability and nitrogen fertility. Onion cultivars vary widely in their pungency levels and in the broad sense heritability of enzymatically produced pyruvic acid content which has been estimated to range from 48% to 53% (Lin *et al.*, 1995).

Onion pungency could be estimated by analyzing enzymatically-produced pyruvic acid (Schwimmer and Weston, 1961). This acid is produced by allinase by the hydrolysis of a group of flavour precursor S-alk(en)yl-L cystein sulfoxides in onion tissue when they are mechanically chopped or macerated. The reaction products are pyruvate, ammonia and volatile sulfur compounds characteristic of onion flavor and aroma (Lancaster and Boland, 1990). A high correlation between enzymatically produced pyruvic acid and pungency perception was reported by Schwimmer and Weston (1961) and Waller and Corgan (1992). Pyruvic acid concentrations were suggested as mildness selection criteria in onion breeding (Waller and Corgan, 1992). Therefore, the pyruvate content developing in homogenized shallot tissue indicates the potential pungency of the bulbs.

2.3. Nitrogen Requirements of Plants

Of the three major plant food elements, nitrogen exerts most noticeable effects on plants (Reiley and Shry, 1979), as it is required in the greatest quantity by most crops. It has also complex behavior, occurring in the soil, air and water in inorganic and organic forms, in which it poses the most difficult problem in making fertilizer recommendations (Archer, 1988).

One of the main problems in determining appropriate amount of nitrogen to the soil is, accounting for all the nitrogen in a soil-plant system. Nitrogen may be lost by leaching through the soil to the drains or aquifer and also in a gaseous form to atmosphere. Most crop plants take up both ammonium and nitrate ions through the root system. Most uptake at normal soil pH levels for crop production occurs as a nitrate due to the rapid conservation of ammonium to nitrate in the soil following application of any ammonical fertilizers (Archer, 1988). Nitrogen also seems to regulate the use of other major elements (Reiley and Shry, 1979). Henry and Raper (1989) observed that nitrogen uptake by plants was a function of root mass and specific rate of uptake.

The main forms in which nitrogen is added to the soil as inorganic fertilizers are nitrate (NO_3), ammonium (NH_4^+) and simple amides (-NH_2). In addition, nitrogen is supplied as organic fertilizers, like animal manure, containing both ammonium and organic forms. Apart from these, nitrogen is also added to the soil as gaseous nitrogen (N_2), being fixed from the atmosphere (Archer, 1988).

Nitrogen has been identified as being the most limiting nutrient in the plant growth. Plants absorb nitrogen in the cation form (NH_4^+) or anionic form (NO_3^-). They obtain readily available N forms from different sources. The major forms include: biological nitrogen fixation by soil microorganisms, mineralization of organic N, industrial fixation of nitrogen gas and fixation as oxides of nitrogen by atmospheric electrical discharge

(Tisdale *et al.*, 1995). The availability of nitrogen through biological N fixation is influenced by the soil pH, and its mineral nutrient status, photosynthesis, climate and crop management (Miller and Donanue, 1995; Tisdale *et al.*, 1995). Similarly, mineralization of organic nitrogen to inorganic forms depends on temperature, level of soil moisture and supply of oxygen (Tisdale *et al.*, 1995).

The available nitrogen form could become unavailable or lost via plant uptake, denitrification, volatilization, leaching, and ammonium fixation (Tisdale *et al.*, 1995). The loss of available nitrogen through natural processes is believed to surpass the gain (Miller and Donanue, 1995; Tisdale *et al.*, 1995). This fact has made fertilizer management an important aspect of crop production practices (Kleinkopf *et al.*, 1987). Consequently, nitrogen is applied relatively in large quantities all over the world (Sanchez, 1976; Miller and Donanue, 1995). The deficiency of nitrogen has an overriding control on plant growth and dominates the effect of other plant nutrients. The deficiency symptoms of nitrogen in plants generally include stunted plant growth, spindly appearance of the plants, reduced growth of leaves, chlorosis and premature senescence of older leaves and restricted root growth and branching (Marschner, 1995; Miller and Donanue, 1995; Tisdale *et al.*, 1995).

Nitrogen plays a central role in plant biochemistry as an essential constitute of cytoplasmic proteins, nucleic acids, chlorophyll, cell walls and a vast array of other cell components. Consequently, a deficiency in the supply of nitrogen has a profound influence upon crop growth and can lead to a total loss of yield in extreme cases (Hay and Walker, 1989; Marschner, 1995). Numerous studies have shown the limiting effects of N deficiency on the growth and development of crop plants (Muchow, 1988; Muchow and Davis, 1988; Mc Cullough *et al.*, 1994; Muchow, 1994). Muchow (1988) and Mc Cullough *et al.* (1994) observed reduced plant growth and development arising from reduced leaf emergence rate and leaf area development. The adverse effects of N deficiency on plant height, plant N uptake, leaf area index, leaf area duration, crop photosynthetic rate, radiation interception, radiation use efficiency are well documented in the review of Novoa and Loomis (1981) and the work of others (Muchow, 1988; Muchow and Davis, 1988). Biomass production, which is largely dependent on leaf area index, is also strongly dependant on leaf N (Muchow and Davis, 1988).

Nitrogen fertilizer is essential to increase crop yield, crop quality and production efficiency as it is the nutrient absorbed in the greatest amounts by plants in terms of equivalents (Kafkafi and Genbaum, 1971). They also noted that in most crop systems, available nitrogen is often a more limiting factor for the plant growth than any other nutrients. The strategy of maximizing crop yield by supplying high levels of fertilizer to the soils, without regard to the efficiency of utilization, is being questioned because of high costs of fertilizers and increasing concerns over environmental pollution (Konesky *et al.*, 1989).

Thus, improving crop production and productivity requires the use of nitrogen fertilizer with great emphasis on the efficiency of nitrogen utilization. Considering the high cost and the detrimental effects and nitrogen deficiencies on crop production the efficient use of nitrogen in crop production has become a desirable agronomic, economic, and environmental goal (Le Gouis *et al.*, 2000). Nitrogen use efficiency could be improved through improved agronomic practices and through growing cultivars efficient in nutrient use (Epstein and Jefferies, 1964).

2.4. Nitrogen Fertilizer Requirements of Onions and Shallots

Three major essential plant nutrients, nitrogen, phosphorus and potassium are known to be increasingly in short supply in the soils of Eastern, Western and Southern Africa (Rao *et al.*, 1998). Particularly nitrogen and phosphorus are deficient in many soils of tropical Africa (Richardson, 1968) which might also be true for many Ethiopian soils (Murphy, 1959 and Murphy, 1968).

Considering the status of the soil, additional N application may be necessary in order to meet the crop N requirements. The amount of nitrogen needed is usually based on soil organic matter content, crop uptake and yield levels. Nitrogen uptake levels by onion crops may vary from less than 50 kg to more than 300 kg ha^{-1}, depending on cultivar, climate, plant density, fertilization and yield levels (Hegde, 1986, 1988; Sørensen, 1996; Suojala *et al.*, 1998; Pire *et al.*, 2001).

Onion is a heavy feeder, requiring ample supplies of nitrogen. Too much nitrogen can result in excessive vegetative growth, delayed maturity, increased susceptibility to

10

diseases, reduced dry matter contents and storability and thus, result in reduced yield and quality of marketable bulbs (Brewster, 1994; Sørensen and Grevsen, 2001). On the other hand, under sub-optimal supply of N, onions and shallots can be severely stunted, with bulb size and marketable yields reduced. Hence, sub-optimal levels of this nutrient in the soil adversely affect the yield, quality and storability of bulbs of onions and shallots (Brewster, 1994; Gubb and Tavish, 2002).

Onion bulb size is related to planting density where smaller bulbs are formed at closer spacing. However, size can be increased by application of nitrogen and potash during growing period (Rice et al., 1993), while Kebede et al. (2002a) reported that the addition of potassium had no effect on the yield of shallot.

According to Kebede et al. (2002b), nitrogen fertilization could reduce yield of rain-fed shallots. Both in the main and in the short rain seasons, soil moisture contents were found to be low during peaks of the growing periods. With increased N rates, shallot plants tended to have more vegetative growth, delayed maturity and reduced bulblet sizes. When crops received supplemental irrigation, however, more marketable bulbs per plant were obtained with increased rates of N, which in turn raised yields up to fertilization rates of 150 kg N ha^{-1}. Similarly N utilization in the onion crop was shown to correlate with the availability of soil moisture (Hegde, 1986, 1988; Wiedenfeld, 1994).

On the contrary, the yields of irrigated shallot crops during dry and warm season were little affected by N fertilization (Kebede et al., 2002a, 2003a). This could be due to warmer temperature, which enhances release and availability of N to plants compared to relatively cooler temperature during the rainy season (Marschner, 1995). Moreover, the response of shallots to applied N during the rainy seasons could also be due to N loss by leaching and competition from higher weed populations compared to the dry season irrigated crops (Kebede et al., 2002b).

Increasing N rates did not significantly affect lateral branching in shallot but improved the dry matter, soluble solids and pungency of the bulbs. However, improvement in these parameters may be of little significance for growers who are not selling their produce on quality basis. Moreover, a higher risk of bulb storage losses was observed in N fertilized plants (Kebede et al., 2002a, 2003a). Batal et al. (1994) also showed that high levels of

11

nitrogen fertilization promote sprouting and decay of onions. Hence, N fertilization practices by growers need reconsideration on the basis of soil fertility and soil availability, growing season and the possibility of better storage facilities or immediate use of the bulbs.

Neeteson *et al.* (1999) indicated that efficient nitrogen fertilization should consider the risk of leaching losses of the nitrate and possible damage to the environment. Large quantities of N fertilizers may be required to obtain high yields of quality onion bulbs, but large amount of N may also remain in the soil after harvest. In the Netherlands, De Visser (1998) estimated N leaching which was about 50% of the 100-120 kg ha^{-1} N commonly applied to onion fields. In Japan, Hayashi and Hatano (1999) also calculated that the N leached annually from an onion field could correspond to 58% of applied N. Generally, the amount of N applied and onion crop responses vary from place to place. High yielding varieties usually require more nitrogen than low yielding ones. Results from different climatic regions of the world also show varying responses of onions to applied nitrogen. Hence, assessment of the right amount of nitrogen fertilizer applied is essential in minimizing the risk of nutrient loss in addition to economic considerations.

2.5. Storage of Onions and Shallots

Bulbs are natural storage organs well adapted for long-term crop storage. Careful handling and the choice of a suitable storage method for the cultivar type in question are vital to ensure that the product retains its quality until it reaches the consumer, retaining an attractive appearance (Gubb and Tavish, 2002). The storage structures can influence the post-harvest deterioration of onions and shallots. Storage structures around the world vary from simple shelters to insulated buildings with automatically controlled heating and cooling (Uzo and Currah, 1990).

At small scale level, simple, naturally ventilated storage structures constructed from poles, wire meshes and grass-thatched roofing oriented to the direction of prevailing wind, can effectively be used for extending the shelf life of onion. Racks or tiers having two or three layers of bulbs would be desirable for proper storage (Gubb and Tavish, 2002). The most important consideration here is that onions should be thoroughly matured, cured and dried before storage. Farmers in some areas of our country store shallots on a raised wooden bed where alternate piles of bulbs and wheat straw are made at a ratio of 2:1 (two piles of shallot bulbs to one layer of wheat straw between the piles) (Getachew and Asfaw, 2000).

Bulbs can be harvested with tops which are sometimes tied or plaited together into bundles and then hung up. This is done in Japan, where onions are stored in bundles hung on bamboo poles in ventilated shelters to protect them from the sun and rain (Thompson *et al.*, 1972). In India and the Sudan, onions are bulk-stored in special houses with conical thatched roofs where the walls are made of bamboo canes which are well spaced to ensure air circulation (Sulafa *et al.*, 1973). The bulbs can also be stored in crates or bags in well-ventilated houses. The crates are stacked on slatted floors in storage houses built so that air can circulate from below the floor, upward through the crates and out through vents in the roof (Uzo and Currah, 1990).

There is the danger of excessive neck rot developing when onions are stored with green tops. Hence, Ali and Shabrawy (1979) suggested that bulbs harvested with their tops should be cut to a 1 or 2 cm long top for best storage result. It has also been reported that delayed harvesting increased sprouting in store (Ward, 1979) though in Egypt a decrease in disease severity during storage has been found with delayed harvesting (Ali and Shabrawy, 1979).

Poor handling of onions before or during storage can increase bulb losses. Damage due to poor handling is most commonly manifested as bruises caused during harvest. The bruises cause increased respiration rates, and infestation with diseases and pests, thus accelerating weight loss (Green, 1972). Therefore, bulbs intended for storage must be cured before topping, must be as free as possible from cuts and bruises and must be handled with maximum care. Major post-harvest losses in shallots and onions are mostly due to sprouting and rot (Green, 1972; Sebsebe, 2006).

According to Uzo and Currah (1990), the bulbs should first be well cured, then stored at low temperatures around 0 °C; or, if not available, high temperatures of about 30 °C with relative humidity of between 65 to 80 % should be used. This will help to reduce losses during storage.

Dry shallot bulbs are sold either fresh or from storage. Shallot clones vary considerably in storage life, with a range of 2 to 9 months (Currah and Proctor, 1990), and storage temperature and genetic traits are the main factors that influence storage life (Jones and Mann, 1963; Currah and Proctor, 1990; Brewster, 1994; Grubben, 1994). Also Currah and Proctor (1990) and Grubben (1994) showed that shallots could be stored for long periods under ambient conditions in the tropics, over 5 months in some trials. Storage in shaded heaps in the field or in open sheds under ambient conditions is common in the tropics (Getahun et al., 2007).

Storage extends the availability of bulbs over long periods. Storage disadvantages are dry matter and moisture losses. Following dormancy breaking, they normally resume growth and loose their food value. The onset of dormancy is thought to be caused by translocation of growth inhibitory substances from the leaves to the bulbs as the crops mature (Komochi, 1990). Stow (1976) found inhibitors in the leaves of onions approaching maturity and showed that defoliation at this stage shortened dormancy. Abscissic acid was identified but was accounted for only 10-20 % of the growth inhibitory activity.

During subsequent storage periods, Isenberg et al. (1974) reported a progressive decline in inhibitory activity extractable from bulbs, followed by increase first in cytokines activity and then by gibberellins and auxin activity. Other possible losses include decay, sprouting, and rooting (Rubatzky and Yamagunchi, 1997). The later maturing, pungent, globe type onions, which tend to have more bulb dry matter, can be stored and can be held for no longer than 2 months; however, these types are usually consumed soon after harvest. Bulbs started from sets generally do not store well and should be used soon after harvest (John, 1992).

During storage, translocations of carbohydrates occur via the stem plate from the outermost succulent swollen scales to inner scales. The outermost succulent scale gradually desiccates, becoming a dry protective scale that helps to reduce water loss from inner succulent scales. This process can continue, resulting in an increase in the number of dry outer scales and in turn, a decrease of an equal number of succulent scales, along with a concomitant decrease in bulb diameter (Rubatzky and Yamagunchi, 1997).

Onions respiration rate are generally low, but do increase with elevated temperature. Relative humidity has a large influence on storage life; sometimes its influence is greater than that of temperature. While the optimum range of relative humidity is 65-75 %, favorable temperature regimes are either low (0-5 °C) or high temperature (25-30 °C) (Gubb and Tavish, 2002). In the tropics, in the absence of refrigerated stores, the storage of onions at 25 °C within the range of 50-70 % RH produces the least spoilage (Mondal and Pramanik, 1992). Respiration rate is related exponentially to increased storage temperature between 0 and 20 °C and is generally a good indication of postharvest quality degradation (Peiris *et al.*, 1997). Respiration of damaged bulbs is more rapid than that of intact ones and this can result in higher water-vapor production in the storage environment if not controlled by ventilation, which can lead to rooting and then to sprouting (Gubb and Tavish, 2002).

Onions are suited for extended storage relative to other vegetable crops. However, both pre and postharvest conditions could affect quality and storability during postharvest life (Jones and Mann, 1963). John (1992) reported the range of relative humidity to be lower for onion storage than for most vegetables because dampness in storage causes considerable rots and mould growth. High temperature and moisture induce new growth and if the crop can be kept cool and dry, and well ventilated, it can be held in common storage for several months. Forced-air ventilation improves the removal of excessive humidity and heat. It helps to keep the outside layers of the onions dry and the bulbs dormant.

The essentials of successful storage are thorough ventilation, a uniform and comparatively low temperature, low humidity, proper maturity and freedom from disease infection. *Aspergillus fumigatus* and *Penicillium spp.* frequently occur in the microflora of stored temperate onions, but the former flourish only at >40 °C and the later at 1-5 °C or 20-25 °C (Hayden and Maude, 1997). From stored onions in Yemen, Maude *et al.* (1991) isolated eleven distinct bacterial or yeast organisms, several of which were also human pathogens or which live in the gut (e.g. *Pseudomonas aeruginosa, Entercoccus faecalis*); many of them were found in combination in the rotting bulbs. They concluded that, in the prevailing high temperatures senescent onion tissues were likely to be invaded by a wide range of opportunistic organisms, which speed up the break-down of the drying bulb scales. Better husbandry practices, including cutting the tops off further form the flesh of the necks, were advised.

The proneness of short day onions to post-harvest deterioration is due to both genetic factors and the technology employed in the storage of the bulbs. Bendarz *et al.* (1986), in their trials on yield and quality of various short day onion varieties in Northern Nigeria, found that Texas Early Grano, Yellow Bermuda, Red Creole, and F_1 Granex cultivars were resistant to bolting and gave high yields but suffered high storage losses as compared with local cultivars. Storage losses in 37 onion cultivars in India were found to be positively correlated with the bulb protein content and negatively correlated with ash, potassium, dry matter, total soluble solids, and nonreducing sugar content. Another Indian study showed that the high pungency dry season cultivar Ropali was suited for storage dehydration on the basis of its high dry matter and insoluble solids content (Maini *et al.*, 1984).

Uzo and Currah (1990) reported that the size of onion bulbs affected both sprouting and water loss during storage. They found that in storage at 11°C, larger bulbs sprouted at a faster rate than smaller ones, but that small onions lose weight more rapidly. Saimbhi and Randhawa (1982) found that storage losses (rotting) were greatest in larger bulbs and least in smaller bulbs. In contrast, Bielinska- Czarnecka *et al.* (1981) reported that the smallest bulbs were the first to sprout and produce roots and that the largest bulbs were the last. However, in some other investigations, no effect of bulb size on date of commencement of sprouting was found (Kato, 1966; Ward, 1979).

Storage temperature and humidity affect sprouting and rooting, loss in weight, respiration rate, the incidence and severity of rots and many other qualities of stored bulbs. Selection of suitable storage temperature and humidity can preserve the bulbs for months (Uzo and Currah, 1990). Loss in weight of bulbs increases with increasing storage temperature (Hurst *et al.*, 1985). Uzo and Currah (1990) reported an increase in weight losses of bulbs with in increase in temperature from 0 to 10 °C followed by a decrease in losses up to 27 °C and further increase at still higher temperatures. This effect is probably as a result of an increase in sprouting between 0 and 15 °C (Hurst *et al.*, 1985) combined with an increasing respiration of onions at higher temperatures (Thompson *et al.*, 1972). However, the respiration rate of onion rises more slowly as temperature rises than is the case with other crops (Lutz and Hardenburg, 1968). This may contribute, in part, to the feasibility of storing onion bulbs at high temperatures. Yamaguchi *et al.* (1957) found a similar percentage of sound bulbs resulting from storage at 0 to 7.5 °C to that from 25 to 30 °C after 4 months of storage. At temperatures between 7.5 and 25.5 °C they recorded a significant reduction in the percentage of sound bulbs.

2.5.1. Effect of nitrogen on shelf life of onions and shallots

Various cultural practices influence storage ability of onion (Thompson *et al.*, 1972). Brewster (1994) described a web of complex interactions among factors contributing to quality of bulbs in post-harvest storage; these include mineral nutrition, cultivar, stage of bulb development/maturity, conditions during maturation and harvesting and curing. Timing and types of various fertilizer applications have been reported to have effects on disease incidence, weight loss, and re-growth in storage (Proctor *et al.*, 1981). Vaughan (1960) noted an increased incidence of disease particularly neck rot *Botrytis allii* following excessive nitrogen application and late irrigation. Jones and Mann (1963), Currah and Proctor (1990) and Brewster (1994) indicated that excessive application of nitrogen in the growing season results in the production of bulbs with thick necks which do not store well. In India, however, Wayse (1967) found that in addition to yield increases, a low nitrogen application (45 kg ha^{-1}) significantly reduced weight losses in storage but had a variable effect on sprouting.

Brewster (1994) described a web of complex interactions including N fertilization among factors contributing to quality of bulbs in post-harvest storage. Accordingly, many researchers found that high levels of nitrogenous fertilizer resulted in reduced onion storage life (Wayse, 1967; Proctor *et al.*, 1981; Batal *et al.*, 1994), though others produced differing results, perhaps dependent on the requirements of specific cultivars. For instance, Zafrir (1992) demonstrated that biweekly applications of N throughout the growing season, amounting up to 500 kg ha^{-1} at final application, had no adverse effect on quality and keeping ability of long storing onion cultivars.

Adequate N supply promotes rapid and complete bulb maturity, which is essential for good storage. It is known that reasonably high levels of nitrogenous fertilizers result in reduced onion storage life (Kato *et al.*, 1987). However, bulb from plants with higher tissue N level often are of poor quality and can not be stored well for longer periods

17

(Henriksen, 1987). Jones and Mann (1963) also concluded that excessive application of N during production results in bulbs with short shelf-life during storage. However, the levels of N that adversely affect storability of bulbs vary with species and cultivars.

Sprouting is the major factor limiting storage life of onion and shallot bulbs. At harvest, bulbs are in a state of innate dormancy and dormancy terminates when inner sprout growth begins (Brewster, 1987). High dose of N produces quick sprouting of thick-necked bulbs during storage. Moreover, greater percentage of open thick-necked bulbs results in increased sprouting due to increased access of oxygen and moisture to the central growing point. Nitrogen above 100 kg ha^{-1} increased the sprouting of onion bulbs both under normal and cold storage conditions (Dankhar and Singh, 1991). According to the authors sprouting percentage of onion bulbs was increased with the increase in N application rates.

Sebsebe (2006) found that application of N fertilization increases the percentage of sprouted shallot bulbs in comparison with low N application. Similarly, Bhalekar et al. (1987) reported 6.7, 11.6, and 13.1 % sprouting in onion plants fertilized with 0, 75, and 150 kg N ha^{-1}, respectively, during periods under common storage condition. Celestino (1961) found bulbs fertilized with 60 or 120 kg N ha^{-1} sprouted twice as much under common storage compared to those which were not fertilized. However, application of N had no significant effect on the percentage of sprouted bulb under cold storage condition. Another factor that could be attributed to the increment in sprouting of bulbs is higher concentration of growth promoters than inhibitors in the bulbs of N fertilized plants that keep it growing (Dankhar and Singh, 1991).

Onion bulbs produced without N application resulted in lowest rotting (22 %), while highest rotting (36 to 54 %) was recorded in bulbs produced under higher dose of N (Jones and Mann, 1963). Report by Celestino (1961) also shows N application increased rotting of bulbs under both common and cold storages.

Dankhar and Singh (1991) reported that weight loss of bulbs increased with the increase in the N level. Some cultivars lose the skin of the bulb relatively easily through cracking. The loss of skins detracts the appearance of bulbs and lowers their commercial value (Komochi, 1990).

Therefore, it is imperative that good cultural practices are used for onions and shallots, which include optimum nitrogen fertilization and avoiding late irrigation of production fields. Breeding of bulbs with thin necks, and harvesting of ripened firm bulbs with entire skins should be also encouraged whenever harvesting is carried out (Uzo and Currah, 1990). These will help reduce storage losses of onions.

2.5.2. Cultivar differences in shelf life of onions and shallots

The commercial storage period for onions is often terminated by the initiation of root growth and sprout elongation. Immediately after harvest, the bulbs are in a natural state of rest that is controlled by endogenous hormone levels and varies with genetic makeup of the particular cultivar (Currah and Proctor, 1990).

Adequate metabolic substrate levels (primarily carbohydrate but may also include organic acid) are essential for storage of any crop. In onions, fructose level at harvest has been suggested as an indicator of storage potential (Rutherford and Whittle, 1984). Gorin and Borcsok (1980) considered total sugar content during storage to be an index of keeping quality. Carbohydrate in onion bulbs account for a major portion of their dry weight and include fructose, glucose, sucrose, a series of oligosaccharides, and possibly arabinose and ribose (Rutherford and Whittle, 1982). Organic acids that have been identified include pyruvic, malic, citric, fumaric, and α-ketoglutaric (Take and Otsuka, 1967).

Onion bulbs contain high concentration of fructans, which constitute a major portion of water soluble carbohydrates and have been considered to be associated with storage life of the bulbs (Darbyshire and Henry, 1981; Rutherford and Whittle, 1982, 1984). Fructans are a series of fructosyl polymers based on sucrose, with varying degrees of polymerization. Rutherford and Whittle (1982) studied changes in the carbohydrate composition of onion during storage and reported that the main change was the hydrolysis of oligosaccharides to reducing sugars.

According to Kebede et al. (2003a), frequent irrigation (75% available soil moisture) was not only favourable for maximizing yields of shallot but also reduced the quality and storability of the bulbs in 'Dz-sht-91' shallot variety. Hence they suggested that this cultivar should be irrigated frequently only if it is to be consumed immediately after harvest or if alternative storage other than ambient conditions is available. The authors noted that application of nitrogen didn't have clear effect on yield, quality and storability of both local (Fedis) and improved (Dz-sht-91) shallot cultivars.

Several researchers also reported cultivar differences in storability of onion bulbs. For instance, Stevenson and Cutcliffe (1982) reported onions produced in Prince Edward Island, Canada, were stored in a conventional storage room for 3 to 4 months, depending on cultivars. Hurst *et al.* (1985) reported that white onions showed higher losses than yellow skinned ones. Saxena *et al.* (1974) also reported that red cultivars have higher storage potential in comparison with yellow and white cultivars under conditions of Guyana.

High storage losses compel shallot producers in Ethiopia to sell their produce immediately after harvest when the price is low. At harvest time, the market is over-flooded and the price is very low, while the supply is low and the price is high at other times. Absence of cultivars with good keeping quality and improved storage facilities usually aggravate this problem. Many farmers in Ethiopia are unable to keep planting materials from their own shallot harvest for the next planting season and they often buy from other areas producing shallot bulbs under irrigation, which are expensive. Working on fifty-seven local cultivars, only few cultivars retaining 60-67 % marketable bulbs were obtained after four months of storage at Debre Zeit under ambient conditions (Getahun *et al.*, 2007). For most of the cultivars, they observed storage losses greater than 60% during the same storage period. However, there is no information on the storage potential of the local shallot genotypes produced under different management practices. Therefore, it is important to be acquainted with the storability of those promising and recommended shallot cultivars produced under different rates of N fertilization.

3. MATERIALS AND METHODS

3.1. The Experimental Site

The field experiment was conducted during the dry season of 2007/08 at Sirinka Agricultural Research Center, Haik/Jari sub-center, which is located at 437 km Northeast of Addis Ababa. Haik Agricultural Research Sub-center is located at $11°21'$ N latitude and $39°38'$ E longitude and at an altitude of 1680 m above sea level. The mean annual rainfall is 1204.6 mm and average annual minimum and maximum temperatures are 11.2 °C and 25.6 °C, respectively. The experiment was conducted on a silt-sandy soil with a pH of 7.12, organic carbon content of 1.745%, CEC of 32.04 meq/100 gm, total nitrogen content of 0.165%, available P of 8.73 ppm, Ca of 28.22 meq/100 gm, K of 2.25 meq/100 gm and Na of 4.17 meq/100 gm (Appendix Table 1). Soil chemical analysis was conducted in soil laboratory of Sirinka Agricultural Research Center following standard procedures.

The analysis of the collected sample for selected soil properties indicated that the soil of the field experimental site is a silt-sandy and neutral in reaction. The soil is low in total nitrogen (Bremner and Mulvancy, 1982), deficient in available phosphorus (Olsen *et al.*, 1954), very low in organic matter contents (Walkley and Black, 1934). According to Black (1965) the soil of the experimental site had high cation exchange capacity and exchangeable bases (Ca^{2+}, K^+ and Na^+) (Appendix Table 1).

The storage experiment was conducted at Sirinka Agricultural Research Center (located 63 km North of the field experimental site, Haik sub-center). Sirinka Agricultural Research Center is located at $11°83'$ N latitude and $39°68'$ E longitude and at an altitude of 1850 m above sea level. The mean annual relative humidity is 66.5% with a mean annual maximum and minimum temperatures of 26 °C and 13 °C, respectively (Fasika, 2004).

3.2. Treatments and Experimental Design

3.2.1. Field experiment

The treatments consisted of a factorial combinations of four nitrogen (N) fertilizer levels (0, 50, 100 and 150 kg N ha^{-1}) and four shallot genotypes (Huruta, Negelle, DZ-sht-68 and

21

Local). Urea was used as a source of N. Huruta and Negelle are varieties released by the Debre Zeit Agricultural Research Center while DZ-sht-68 is on a pipeline for release by Sirinka Agricultural Research Center (SARC, 2007). The local is a yellow skinned cultivar cultivated by Sirinka and Merto farmers for their pungent flavor that persist after cooking as compared to released and promising red skinned cultivars.

The experiment was conducted in a randomized complete block design with three replications. The plot size was 3 m × 3.2 m (9.6 m^2). Bulbs were planted in rows that were 3 m long at a spacing of 0.4 m between rows and 0.2 m between plants. An alley was left between the blocks and plots, measuring 2 m between the blocks and 1.5 m between the plots within a block. Nitrogen fertilizer was applied in two splits where the first half was applied at 50 % bulb sprout and the remaining half side-dressed one month later. All plots received phosphorous at the rate of 92 kg ha^{-1} as P_2O_5 (Lemma and Shimels, 2003), which was incorporated along the rows at the time of planting. The fertilizer materials used were Urea (46 % N) and triple super-phosphate [TSP (46 % P_2O_5)] as sources of N and P nutrients, respectively.

Plots were prepared by ploughing the soil to 35 cm depth. After harrowing and leveling, ridges and furrows were made. Planting of bulbs was done on December 20, 2007. The plots were irrigated at an interval of five days till the shallot crop matured. Cultural practices such as weeding, pest control and cultivation were uniformly performed in all plots as per the recommendations following crop establishment (Getahun et al., 2007). Harvesting was done on 18th April 2008. After harvesting, the shallot bulbs were cured by windrowing, spreading them thinly on the ground for 7 days before topping and collecting data on bulb yield and yield components.

3.2.2. Storage experiment

Bulbs were harvested from the central six rows of the field grown plants (4 genotypes × 4 N fertilizer levels) and the cured marketable bulbs used for the storage experiment. The storage experiment was began on the 25th of April 2008 and terminated on the 24th of July 2008. Five kg of shallot bulbs were stored from each treatment on shelves in a grass thatched roofing store at ambient atmospheric conditions. At the start of the study the fresh weight of the bulbs was recorded and stored on shelves in three replications using randomized complete block design. Daily storage temperature and relative humidity were recorded at every three hours interval using a hygrotherm. The average daily maximum and minimum temperatures of the storage house during the three months period were 31.6 °C and 15.8 °C, respectively, and the average daily relative humidity was 46% (Appendix Fig. 1).

3.3. Collection of Experimental Data

3.3.1. Field data

Data on growth and yield related traits were recorded from ten randomly taken plants from the central six rows of each plot and the calculated values were used for statistical analysis. Plant height was measured in cm from the ground level to the tip of mature leaf. Total number of leaves and total number of lateral growths/shoots per plant were recorded at physiological maturity. The diameter of the longest leaf was measured in mm at maturity using vernier caliper.

The total number of bulblets per plant was recorded after harvest. The average size of bulbs was measured in mm using caliper at the widest point (middle portion) of matured bulb. Mean bulb weight was recorded as the average weight of matured bulb splits of the sample plants.

Total biomass per plant was recorded as sum of the total bulb yield, above ground parts and roots at the time of maturity and expressed in kg. Harvest index per plant was expressed as the ratio of total bulb yield per plant to the total biomass in percentage.

For the determination of percent dry matter a homogenate was prepared from 10 sample bulbs from each plot and then 250 g of the homogenate was taken and oven dried (Wagtechn Gp/120/SS/100/DIG oven) at a temperature of 70 °C for 60 hours till constant weight recorded. Then the weight was measured using digital balance (METTLER TOLEDO) and percent dry matter was calculated using the formula:

23

$$DW\ (\%) = \frac{[(DW + CW) - CW]}{[(FW + CW) - CW]} \times 100$$

Where: DW= dry weight

CW= container weight

FW= fresh weight

Data on maturity and yield were recorded from the central six rows of each plot. Days to maturity was taken as the actual number of days from emergence to the time when 75 percent of the plants' foliage fell down or senesced. Marketable bulb yield was recorded based on the size/diameter and damage of the bulbs after curing and topping of the shallot crop. Marketable bulbs were graded by diameter into small, medium, and large (20-35, >35-50, >50 mm, respectively) as described by Kebede *et al.* (2003a). Damaged and small bulbs less than 20 mm diameter were recorded as unmarketable bulb yield.

Total soluble solid (TSS) was determined using the procedures described by Waskar *et al.* (1999). Aliquot juice was extracted using a juice extractor and 50 ml of the slurry centrifuged for 15 minutes. The TSS was determined by using a hand refractometer (ATAGO TC-1E) with a range of 0 to 32° Brix and resolutions of 0.20° Brix by placing 1 to 2 drops of clear juice on the prism. The prism was washed with distilled water and dried with tissue paper after each measurement before use. The refractometer was standardized against distilled water (0% TSS).

The content of pyruvic acid developing in homogenized bulb tissue was used as a measure of pungency following the procedure of Ketter and Randle (1998). A core sample was cut from shallot bulbs of each treatment and squeezed using a juice extractor (6001X U.S.A model) and 0.5 ml of the juice was put into a 40 ml test tube. The slurry was allowed to wait for 10 minutes. A 1.5 ml of 5 % trichloroacetic acid was added to each test tube and vortexed; to each test tube, 18 ml of deionised water was added, which was vortexed and capped. From the solution, 1ml was taken and put in a 20 ml test tube and 1 ml of 2, 4-Dinitrophenylhydrazine and 1 ml of deionised water was added to each test tube and vortexed. The test tubes were then placed in a water bath at 37 °C and allowed to incubate for 10 minutes after which 5 ml of 0.6 N sodium hydroxide was added to each test tube and vortexed. The samples were run on a spectrophotometer (JENWAY M6300) set at 420 nanometers and pungency readings were expressed as μ mol ml^{-1} juice. Standards were made and run under the same conditions to prepare a standard curve; then pungency readings on spectrophotometer were determined using the standard curve.

3.3.2. Storage data

In the storage study, the following observations were recorded at a biweekly interval for the three months storage period.

Percentage of weight loss of bulbs was determined using the methods described by Waskar *et al.* (1999). The weight loss data were calculated from 30 bulbs which were randomly taken per treatment and weighed at the beginning and mid of each month. Percentage of weight loss (WL) was calculated using the formula:

$$WL \ (\%) = \frac{Wi - Wf}{Wi} \times 100$$

Where, W_i = initial weight

W_f = final weight

Percentage of bulbs sprouted was cumulative which was based on the number of bulbs sprouted in the biweekly assessment. Data on sprouting was recorded by counting the number of bulbs that sprouted. The sprouted bulbs were discarded after each count to avoid double counting. Bulbs that sprouted and rotted at the same time were classified as sprouted.

Percentage of bulbs rooted was cumulative which was based on the number of bulbs rooted in the biweekly assessment. Data on rooting was recorded by counting the number of bulbs that rooted. The rooted bulbs were labeled after each count to avoid double counting. Bulbs that rooted and rotted at the same time were classified as rooting.

Data on rotting was determined by counting the number of bulbs that rotted. The rotted bulbs were discarded after each count to avoid double counting.

3.4. Soil Analysis

The soil properties described in the study area include soil particle size distribution (texture), pH, organic matter, total N, available P, exchangeable bases (Ca, K and Na) and cation exchange capacity (CEC).

A total of ten sub-samples were collected from the entire experimental field to a depth of 30 cm before planting to determine the different soil physical and chemical properties listed above. The collected surface soil samples were bagged and transported to the laboratory for preparation and analysis of selected soil properties following the standard laboratory analysis methods. In preparation for laboratory analysis, the soil samples were then bulked to make one composite soil sample, air dried, grounded, and made to pass through a 2 mm size sieve for analysis of soil pH, texture, available P, exchangeable bases and CEC; whereas for the determination of organic carbon and total N, the soil samples were made to pass through 0.5 mm sieve.

Soil particle size distribution (texture) was analyzed by the hydrometer method following the procedure described by Day (1965). Soil pH was measured potentiometrically using a pH meter with combined glass electrode in 1: 2.5 soil: water ratio as described by Carter (1993). Organic carbon was determined using the wet oxidation method (Walkley and Black, 1934) where the carbon was oxidized under standard conditions with potassium dichromate in sulfuric acid solution. Finally, the organic matter (OM) content of the soil was calculated by multiplying the percent OC by 1.724. The total N content was determined using the Kjeldahl method by oxidizing the OM with sulfuric acid and converting the N into NH_4^+ as ammonium sulfate (Dewis and Freitans, 1970). Determination of available P was carried out by Olsen method using $NaHCO_3$ as extracting solution (Olsen et al., 1954).

The exchangeable bases (Ca, K and Na) in the soil were determined from the leachate of 1 molar ammonium acetate (NH_4OAc) solution at pH 7.0. Exchangeable Ca was measured by atomic absorption spectrophotometer and K and Na contents were determined using flame photometer (Rowell, 1994). Similarly, CEC was measured after leaching the NH_4OAc extracted soil samples with 10% NaCl solution. The amount of ammonium ion in the percolate was determined by the usual Kjeldahl procedure and reported as CEC (Hesse, 1972).

3.5. Statistical Analysis

The experimental data collected on various parameters were subjected to analysis of variance (GLM procedure) using SAS software program version 8.2. When the analysis of variance indicated the presence of significant differences, mean separation was done using the Least Significant Difference (LSD) test for the nitrogen and genotype main effects and the Duncan's Multiple Range Test (DMRT) for the interaction effects (Gomez and Gomez, 1984). Correlation analysis was carried out by calculating simple linear correlation coefficients between growth, yield and yield components.

4. RESULTS AND DISCUSSION

4.1. Vegetative Growth and Days to Maturity

Nitrogen fertilization very highly significantly ($p<0.001$) increased most of the growth parameters considered in this study. The number of lateral shoots per plant was also increased highly significantly ($p<0.01$) with nitrogen supply (Table 1 and Appendix Table 4). The tallest plant height (124.1%), highest number of leaves per plant (123.8%), largest leaf diameter (123.2%), highest number of lateral shoots per plant (118.6%), highest biomass per plant (162.9%) and latest maturity (10 days delay) were recorded in the plot that received 150 kg N ha^{-1} as compared with the unfertilized control plots that recorded least values of all of the parameters (Table 1). There was an increasing trend in the vegetative growth parameters considered with the increase in the rate of the applied N.

The present result is in agreement with the result of Kebede *et al.* (2003a) who reported that application of nitrogen at the rate of 150 kg ha^{-1} significantly increased the number of leaves and height of shallot plants at Alemaya on a heavy clay soil. In agreement with results of previous works (Hegde, 1986; Maier *et al.*, 1990 and Kebede *et al.*, 2002b), growth of shallot plants responded to N up to the rate of 150 kg ha^{-1}. On the contrary Hussien (1996) reported that nitrogen fertilization did not affect number of leaves per hill and average plant height at 60 days after application of the nutrient on a sandy clay soil at Dire Dawa.

The high rate of nitrogen application resulted in plants with more vegetative growth and bulb setting as compared to the lower rates which could be attributed to the low initial content of the experimental plot. This is corroborated by Hassan and Ayoub (1978) and Brewster (1990, 1994) who suggested that nitrogen increased both vegetative and bulb growth in onions through its effect on cell activities. Nitrogen increases number and size of leaves and also gives dark green color to the leaves. Thus, through increase in photosynthetic activity of the leaves, it further encourages vegetative growth of plants (Archer, 1988; Marschner, 1995).

Table 1. Effect of applied nitrogen rates and genotypes on vegetative growth and days to maturity of shallot

Treatment	Plant height (cm)	Total number of leaves per plant	Leaf diameter (mm)	Number of lateral shoots per plant	Days to maturity	Biomass per plant (kg)
N (kg ha^{-1})						
0 (Control)	49.50^c	58.81^c	5.73^c	4.85^c	101.33^d	0.35^d
50	56.25^b	64.62^{bc}	6.23^{bc}	5.24^{bc}	104.58^c	0.44^c
100	57.67^b	69.46^{ab}	6.38^b	5.43^{ab}	107.33^b	0.51^b
150	61.43^a	72.80^a	7.06^a	5.75^a	110.75^a	0.57^a
Significance	***	***	***	**	***	***
Genotypes						
Huruta	57.84^a	69.93^a	6.62^a	5.12	109.58^b	0.51^a
Negelle	58.43^a	71.20^a	6.64^a	5.29	107.92^c	0.52^a
Dz-sht-68	58.82^a	72.28^a	6.60^a	5.16	111.33^a	0.53^a
Local	49.76^b	52.28^b	5.53^b	5.71	95.17^d	0.31^b
Significance	***	***	***	NS	***	***
SED	1.53	3.13	0.28	0.24	0.68	0.03
CV (%)	6.66	11.54	10.83	10.66	1.56	12.88

Means within a column sharing common letter(s) are not significantly different at p<0.05; ns= non-significant; **, ***= significant at p<0.01 and p<0.001, respectively.

Genotypic differences in plant height, number of leaves, leaf diameter, biomass and days to maturity were also observed to be very highly significant (p<0.001) (Appendix Table 4). The highest plant height, number of leaves per plant, biomass per plant and days to maturity were observed for Dz-sht-68 genotype while the highest leaf diameter and number of lateral shoots were recorded for Negelle and local shallot genotypes, respectively (Table 1). Except for days to maturity, significant differences were not observed among the improved varieties, whereas all the improved genotypes were found to be significantly more in their vegetative growth than the local cultivar.

Unlike the common onions, shallots are distinguished by the production of lateral shoots. However, the genotypes used in this study did not show significant difference in their lateral shoots (branches) while this trait was significantly affected by N fertilization.

Biomass per plant significantly increased with each level of N fertilizer applied. The improved cultivars also showed more biomass per plant as compared with the local cultivar. Biomass yield had positive and significant correlations with plant height (r=0.70***), number of leaves per plant (r=0.78***), leaf diameter (r=0.74***), lateral branches per plant (r=0.33*), and days to maturity (r=0.84***) (Appendix Table 10). This suggests that increment in biomass production was a result of long phenological period, increase in lateral branches, leaves per plant, leaf diameter and overall increase in plant morphology. Fasika (2004) and Abayneh (2001) also reported similar associations of biomass yield with vegetative growth in shallot and onion genotypes, respectively.

4.2. Number of Bulb Splits, Bulb Diameter and Bulb Weight

4.2.1. Number of bulb splits

Nitrogen fertilization and genotype had very highly significant (p<0.001) interaction effects on number of bulb splits per plant (Appendix Table 5). The highest number of bulb splits per plant (13.6) was recorded for the local genotype fertilized with 100 kg N ha^{-1} while the least number of bulb splits per plant was obtained for Huruta, Dz-sht-68, and local genotypes with no N application and for Dz-sht-68 with the application of 50 kg N ha^{-1} (Table 2). For Negelle shallot variety, the number of bulb splits per plant did not show significant difference due to application of N fertilizer while in Dz-sht-68 genotype and the local cultivar number of bulb splits tended to increase with increase in the amount of applied N. On the other hand, application of 50 kg N ha^{-1} significantly increased number of bulb splits of Huruta variety in comparison with the bulb splits observed in the control treatment while successive N rates did not show significant increase in number of bulb splits of the shallot.

The result of N fertilization on Negelle variety is in line with the investigations of Hussien (1996) who reported that application of nitrogen fertilizer did not show a significant effect on the number of bulbs per hill and also with that of Celestino (1961) who reported that the number of bulbs of onions was not affected by nitrogen application. However, results of the rest of the genotypes showed that this trait could be modified by nitrogen fertilization which differ from reports of the above mentioned authors. This result is in line with the observation of many other researchers. In addition to N rates applied, bulb splitting could result from multiple growing points which could be under genetic control (Jones and Mann, 1963; Steer, 1980) and growth in high temperatures and short days (Steer, 1980). Increase in lateral branching resulting in the production of more bulb splits due to N fertilizer have also been reported by Kebede *et al.* (2002a) on irrigated shallots, which is in conformity with the present result.

Table 2. Interaction effects of nitrogen rates and genotypes on number of bulb splits per plant of shallot

Genotypes	N (kg ha^{-1})			
	0	50	100	150
Huruta	7.23^e	9.90^d	9.93^d	10.27^{cd}
Negelle	11.07^{bcd}	11.73^b	11.90^b	12.30^b
Dz-sht-68	8.17^e	8.43^e	11.30^{bc}	11.60^b
Local	7.93^e	11.07^{bcd}	13.60^a	11.87^b
SE±	0.40			
Significance	***			
CV (%)	6.54			

Means followed by the same letter(s) are not significantly different at $p<0.05$; ***= significant at $p<0.001$.
SED(G×N)= 0.56

4.2.2. Bulb diameter

Nitrogen fertilization and genotype had significant ($p<0.05$) interaction effect on bulb diameter (Appendix Table 5). Significantly higher bulb diameter (47.17 mm) was recorded for variety Negelle with the application of 50 kg N ha^{-1}. The least bulb diameter (26.57 mm) was recorded for the local variety with no fertilizer application. Bulb diameter of

Negelle shallot variety fertilized at 50 kg N ha[-1] was about 77.5% higher than those of the unfertilized local cultivar (Table 3). For Huruta and Dz-sht-68, bulb diameter tended to increase with the rate of applied N. For the local cultivar, the highest bulb diameter was recorded at 100 kg N ha[-1], which was about 50.5% increase over those of the unfertilized control plants. This result is in conformity with the observations of Hassan and Ayoub (1978); Hegde (1986); Suojala *et al.* (1998); Hussaini *et al.* (2000) who found that increased N fertilization increased the diameter of onion bulbs.

Table 3. Interaction effects of applied nitrogen rates and genotypes on bulb diameter (mm) of shallot

Genotypes	N (kg ha[-1])			
	0	50	100	150
Huruta	36.47[efg]	39.40[cdef]	44.00[abcd]	44.13[abc]
Negelle	42.20[abcde]	47.17[a]	44.46[abc]	41.90[abcde]
Dz-sht-68	37.50[efg]	38.33[def]	40.27[bcdcf]	45.63[ab]
Local	26.57[i]	35.43[fg]	40.00[hi]	32.53[gh]
SE±	1.75			
Significance	*			
CV (%)	7.79			

Means followed by the same letter(s) are not significantly different at p<0.05; *= significant at p<0.05.
SED(G×N)= 2.48

4.2.3. Mean bulb weight

Nitrogen fertilization and genotype had a significant (p<0.05) interaction effect on mean bulb weight (Appendix Table 5). The highest bulb weights were recorded for Negelle and Dz-sht-68 shallot varieties with the application of 150 kg N ha[-1], which remained at par with those treated at the rate of 100 kg N ha[-1]. For Huruta, 50 kg N ha[-1] resulted in the highest mean bulb weight which was not show significantly different from the bulb weight obtained at higher rate of N. The least mean bulb weight was recorded for the local variety with no fertilizer application (Table 4). This result is in line with the results of Hassan and Ayoub (1978); Hegde (1986); Suojala *et al.* (1998); Hussaini *et al.* (2000); Kebede *et al.* (2003b) who found that increased N fertilization increased the weight of onion bulbs. Similarly, in Germany, Hussien (1996) also observed that application of 100 kg N ha[-1] resulted in significantly high average weight of shallot bulbs as compared to the average weight of bulbs from the control plots. On the contrary, Kebede et al. (2003a) reported absence or reduced response of shallot bulb weight to N fertilization.

Table 4. Interaction effects of applied nitrogen rates and genotypes on mean bulb weight (gm) of shallot

Genotypes	N (kg ha⁻¹)			
	0	50	100	150
Huruta	24.16^{de}	31.54^{ab}	30.74^{abc}	30.43^{abc}
Negelle	28.61^{bc}	29.09^{bc}	30.30^{abc}	34.29^{a}
Dz-sht-68	24.35^{de}	26.50^{cd}	31.86^{ab}	33.45^{a}
Local	12.46^{g}	16.42^{f}	21.93^{e}	20.38^{e}
SE±		1.31		
Significance		*		
CV (%)		8.52		

Means followed by the same letter(s) are not significantly different at p<0.05; *= significant at p<0.05.
SED(G×N)= 1.85

Biomass yield per plant had positive and significant correlations with number of bulb splits (r=0.36*), and bulb diameter (r=0.74***) (Appendix Table 10). This suggests that increment in biomass production was a result of long phenological period, increase in number of bulblets, and bulb diameter. Fasika (2004) and Abayneh (2001) also reported similar results for shallot and onion, respectively. Mean bulb weight also showed a significant correlation (r=0.88***) with biomass yield per plant.

4.3. Bulb Yield and Harvest Index

Marketable and total bulb yields were significantly (p<0.01) affected by the main effects of nitrogen fertilizer and genotypes. However, harvest index was affected neither by the main nor the interaction effects of the two factors (Appendix Table 6). The highest marketable bulb yield (28.31 t ha⁻¹) and total bulb yield (29.31 t ha⁻¹) were obtained from the plots fertilized with 150 kg N ha⁻¹ (Table 5). In this study, higher number of lateral shoots and yield were recorded in N fertilized plots as compared to the control. Similar results were reported on irrigated shallot by Kebede *et al.* (2002a) who noted that any factor that affected the number and size of bulblets developing from the lateral branches had a great impact on yield of the crop.

Application of N at the rate of 100 kg ha^{-1} significantly increased the marketable and total bulb yields by about 14.8% and 15.28%, respectively, as compared to the yield obtained from unfertilized control (22.51 and 23.30 t ha^{-1}). Similarly, application of 150 kg N ha^{-1} increased marketable bulb yield by 25.77% and total bulb yield by 25.82% as compared to the untreated plots. Further increase in N rate above 100 kg N ha^{-1} did not bring about significant changes in yield, suggesting that 100 kg ha^{-1} is the optimum rate to obtain the highest marketable and total bulb yields of shallot at the study area on a silt-sandy soil. The positive yield response could be attributed to the role of N in promoting the growth of the plant and delaying maturity which could have improved assimilate partitioned to the bulbs. Similar results were also reported on onion by Hegde (1986, 1988); Henriksen (1987); Maier *et al.* (1990); Batal *et al.* (1994) and Cizauskas *et al.* (2003) who found that application of 60 to 300 kg N ha^{-1} gave high bulb yield of onion depending on cultivar and climate.

This result is also in agreement with the findings of Celestino (1961); Wayse (1967); Bhalekar *et al.* (1987); Hussien (1996) and Kebede *et al.* (2002a). Kebede *et al.* (2002a) who found that application of nitrogen fertilizer up to 150 kg N ha^{-1} enhanced yields of shallot when the crops received supplemental irrigation. The reports of Hussien (1996) more or less seem to be comparable to this result even though the yield obtained in this study was much lower than that obtained at Hohenheim, Germany. Celestino (1961) also reported that a significant yield increment of onion was recorded at the application of 60 kg N ha^{-1} compared to the yield obtained at the control treatment. However, different results were reported by Hussien (1996) for shallot in Dire Dawa on sandy clay soil and by Brewster *et al.* (1991) for onion, who found that the total bulb yield was not significantly affected by N fertilization. In this study, the increment in shallot bulb yield in response increased nitrogen application appeared to be due to a corresponding increment in number of bulbs per plant, bulb diameter and mean bulb weight.

Table 5. Effect of nitrogen rates and genotypes on marketable and total bulb yields and harvest index of shallot

Treatment	Marketable bulb yield (t ha^{-1})	Total bulb yield (t ha^{-1})	Harvest index per plant
N (kg ha^{-1})			
0 (Control)	22.51^c	23.30^c	62.17
50	24.87^{bc}	25.66^{bc}	57.23
100	25.84^{ab}	26.86^{ab}	55.54
150	28.31^a	29.31^a	54.87
Significance	**	**	NS
Genotypes			
Huruta	27.52^b	28.50^b	57.76
Negelle	29.51^{ab}	30.12^{ab}	57.04
Dz-sht-68	31.34^a	31.91^a	55.31
Local	13.22^c	14.60^c	59.70
Significance	***	***	NS
SED	1.40	1.39	3.21
CV (%)	13.55	12.96	13.70

Means followed by the same letter(s) within a column are not significantly different at p<0.05; ns=non significant, **, ***= significant at p<0.01 and p<0.001, respectively.

The analysis of variance indicated that there were significant (P<0.001) genotypic differences in marketable and total bulb yields (Appendix Table 6). The highest marketable (31.34 t ha^{-1}) and total (31.91 t ha^{-1}) bulb yields were recorded for Dz-sht-68, while the least marketable (13.22 t ha^{-1}) and total (14.60 t ha^{-1}) bulb yields were recorded for the local cultivar (Table 5). The released shallot varieties Huruta and Negelle did not differ statistically although the later had about 7.2% more marketable yield advantage than the former. Similarly, Negelle remained statistically similar to the highest yielding Dz-sht-68 both in marketable and total bulb yields. The observed bulb yield difference among the genotypes could be attributed to genetic differences in the above ground growth such as plant height, leaf number and leaf diameter, which indirectly determine the bulb size through the amount of carbohydrate synthesized and made available for storage in the bulbs. Similarly the differences could be due to genetic variation in their root physiology and development and ability to mobilize and utilize nutrients. This result is in agreement with the results obtained by Suzuki and Cutcliffe (1989) who reported the presence of variability in marketable bulb yield of different onion cultivars. Fasika (2004) also reported a highly significant yield difference among local shallot genotypes.

Total bulb yield showed positive and significant ($p<0.001$) correlations with plant height ($r=0.71$), number of leaves per plant ($r=0.69$), leaf diameter ($r=0.67$), days to maturity ($r=0.90$), bulb diameter ($r=0.71$), mean bulb weight ($r=0.83$), and biomass per plant ($r=0.80$) (Appendix Table 10). These positive and significant correlation of bulb yield with growth characters suggest that N fertilization rates that improve the above ground plant growth and delay maturity improve the physiological capacity of the crop to mobilize and translocate photosynthate to the organs of economic value, which in turn increase the bulb yield. The positive correlation of days to maturity and bulb yield indicated that prolonging days to maturity in shallot could also increase bulb yield. These associations indicate that increased duration of photosynthetic period in response to N fertilization substantially contributed to enhancement of shallot productivity possibly through the production of more assimilates. Similar results were also reported by Nasreen *et al.* (2007) on onion. Bulb diameter had also strong and positive correlation with the total bulb yield of shallot suggesting that the increment of individual bulb size is fundamental to maximizing shallot productivity per unit area.

The positive correlation of total bulb yield with the number of lateral branches per plant and number of bulb splits per plant suggested that genotypes and N fertilization levels producing large numbers of lateral branches could produce larger number of bulb splits thereby increasing bulb yield. Therefore, the findings of this study indicated that shallot genotypes producing relatively large number of bulb splits particularly in response to N fertilization have a high yielding potential.

The positive correlation of total bulb yield per hectare with days to maturity and bulb diameter, suggested that genotypes producing large number of lateral branches could produce large number of bulb splits per plant and thereby increase bulb yield per hectare. The findings of this study indicated that shallot genotypes producing relatively more bulb splits and/or large bulb size could increase bulb yield per hectare. Hence, in most shallot genotypes that produce small sized bulbs, improving bulb size by N fertilization may increase bulb yield.

Significant differences in harvest index were not observed in this study among genotypes (Appendix Table 6). Biological yield per plant showed negative and significant correlations with harvest index per plant (r=-0.36*) (Appendix Table 10), which indicated that biomass yield per plant increased at the expense of bulb yield per plant. This is in conformity with the work of Fasika (2004) and Abayneh (2001) in shallot and onion, respectively.

Nitrogen fertilization and genotypes had very highly significant interaction effect on unmarketable bulb yield of shallot (Appendix Table 6). The highest unmarketable bulb yield (2.16 t ha^{-1}) was recorded for the local genotype fertilized with 100 kg N ha^{-1}. This could be due to increased number of bulb splits resulting in the production of large number of small sized unmarketable bulbs (Table 6). Generally in Huruta and the local cultivar, N fertilization increased the production of unmarketable bulb yields parallel with the increment in total bulb yield. In Negelle and Dz-sht-68, N fertilization did not have significant effect on unmarketable bulb yield.

Table 6. Nitrogen rates, genotypes and their interaction effect on unmarketable bulb yield (t ha^{-1}) of shallot

Genotypes	N (kg ha^{-1})			
	0	50	100	150
Huruta	0.78^{def}	0.94^{bcde}	1.01^{bcd}	1.18^{bc}
Negelle	0.62^{ef}	0.57^{ef}	0.43^{f}	0.81^{cdef}
Dz-sht-68	0.49^{f}	0.59^{ef}	0.45^{f}	0.75^{def}
Local	1.08^{bcd}	1.03^{bcd}	2.16^{a}	1.27^{b}
SE±		0.119		
Significance		***		
CV (%)		23.37		

Means followed by the same letter(s) are not significantly different at p<0.05; ***= significant at p<0.001.
SED(G×N)= 0.17

4.4. Bulb Quality

Nitrogen fertilization affected the yield and growth traits of shallot. However, it did not have any significant effect on bulb quality traits like pungency, bulb dry matter and TSS.

were not observed in this study (Appendix Table 7 and Table 7). The result on dry matter and TSS is in agreement with previous reports on shallots at Dire Dawa (Hussien, 1996) and Alemaya (Kebede *et al.*, 2002a) and on onions (Maier *et al.*, 1990). On the contrary, opposing results on pungency and TSS were reported by Kebede *et al.* (2003a) and Abbey (2004), respectively, who found that the higher rates of N tended to increase the pyruvate concentration and significantly reduced soluble solid contents of bulbs. The work of Hussien (1996) also showed increased level of pyruvate in shallot bulbs with increased application rates of nitrogen, which is in contrast to the present result.

The application of nitrogen fertilizer during the growing seasons did not result in a significant difference in the percent dry matter content of shallot bulbs at harvest (Appendix Table 7 and Table 7). Similar results were also reported by Celestino (1961), Singh (1987) and Hussien (1996). The absence of significant difference in bulb dry matter content to the application of N might be attributed high GA activity which leads to higher carhohydrate allocation to shoots/leaves during N fertilization, while low GA level resulting in more dry matter allocation to the bulbs. The second reason might be due to the production of more foliage at the expense of production of bulbs. This means, under high nitrogen supply, partitioning was affected whereby yield was sink-limited but not source-limited. Therefore, in line with these arguments nitrogen may not be expected to cause any significant difference in the dry matter content of bulbs.

The TSS content of bulbs was not significantly affected by the application of nitrogen fertilizer. The mean value of TSS observed in this study seemed to be lower as compared to that of Hussien (1996).

On the other hand, verly highly significant genotypic differences in bulb dry matter and total soluble solids were observed in this study (Appendix Table 7 and Table 7). Bulb dry matter and TSS of the local cultivar were about 16.2% and 9.6%, respectively, higher than the improved shallot genotypes (Huruta, Negelle and Dz-sht-68). In conformity with this finding, Fasika (2004) reported that significant differences were observed in dry matter content of local collection of shallot genotypes. Similarly, varietal differences in bulb dry matter content of improved (Huruta) variety and local (Fedis) cultivar was reported by Kebede *et al.* (2003a). Differences in pyruvate contents between the genotypes were not significant. This result is that is in line with the results of Kebede *et al.* (2003a) who reported that there was no variation in pungency between local cultivar (Fedis) and Dz-sht-91 genotype.

Table 7. Effect of nitrogen fertilization and shallot genotypes on the dry matter, pyruvate content and total soluble solids of bulbs

Treatment	Bulb dry matter (%)	Pyruvate content (μ.mol.ml^{-1} juice)	TSS (%)
N (kg ha^{-1})			
0 (Control)	17.20	7.27	13.25
50	16.69	7.25	13.82
100	16.67	7.41	13.68
150	16.37	6.96	14.30
Significance	NS	NS	NS
Genotypes			
Huruta	16.24[b]	7.01	13.68[b]
Negelle	16.54[b]	7.14	13.58[b]
Dz-sht-68	15.47[b]	6.82	13.25[b]
Local	18.68[a]	7.91	14.80[a]
Significance	***	NS	***
SED	0.42	0.42	0.34
CV (%)	6.12	14.46	6.06

Means followed by the same letter within a column are not significantly different at p<0.05; ns=non significant, ***= significant at p<0.001.

Total bulb yield had a negative and significant correlation with pungency (r=-0.40), TSS (r=-0.40) and dry matter content (r=-0.61) of the bulb (Appendix Table 10). The result of this study indicated that genotypes and N fertilization producing high total bulb yield could decrease the shallot bulb quality. Pungency showed slight positive correlation with dry matter content (r=0.31) and TSS (r=0.26) of the bulbs. This result is in contrast to that of Fasika (2004) who reported that genotypes that had high TSS content were less pungent.

4.5. Storage of Shallot Bulbs

4.5.1. Percent weight loss

Nitrogen fertilization and genotype had significant interaction effect on percent weight loss of bulbs only in the last biweekly storage assessment period (Appendix Table 8). There was a significant (p<0.05) cumulative percent weight loss of bulbs during the 4th and 5th biweeks of storage period due to nitrogen fertilization (Table 8). This may be due to increased transpiration, as percent dry matter may have decreased with increased nitrogen supply, which is also supported by reports of Celestino (1961), Singh (1987), Bhalekar et al. (1987) and Hussien (1996).

There was a highly significant effect on cumulative percent weight loss in the 4th (p<0.01) and 5th (p<0.001) biweeks of storage due to genotypic differences (Table 8). A large percent of weight loss was observed in the bulbs of the local cultivar with yellow skinned color as compared to the rest of genotypes with red skinned bulbs. The reason for high weight loss in the local cultivar might be due its thin outer skin that was cracking and peeling off and exposing the underneath succulent tissue to moisture losses. The second reason might be due to the size of the bulbs in which the small sized bulbs have more surface area to volume ratio exposed to the external atmosphere compared to those large sized bulbs of the improved cultivars thereby increasing the percentage of weight loss. After 75 days of storage, the bulbs cumulative weight loss ranged from 48.95% to 62.72% in the Dz-sht-68 and local cultivar, respectively. This result is in conformity with the work of Saxena et al. (1974) who reported that red skinned cultivars have a higher storage potential in comparison with yellow and white ones under conditions of Guyana.

The application of nitrogen fertilizer did not have any significant effect on the total weight loss of bulbs during the early stage of storage (Table 8). Starting from the 4th biweek up to the end of the storage period, however, bulbs grown under 100 and 150 kg N ha^{-1} had significantly more weight loss compared to those from unfertilized plots. Nitrogen fertilization at 150 kg ha^{-1} recorded bulb weight loss of about 17% and 15% over the control and N rate of 50 kg ha^{-1} after storage periods of 4th and 5th biweeks, respectively. In general, it seemed that the total cumulative percent weight loss increased in all genotypes with the increase in the amount of nitrogen applied.

Table 8. Effect of nitrogen fertilization and genotypes on cumulative percent weight loss of marketable bulbs of shallot in ambient storage at Sirinka

Treatment	Cumulative percent weight loss after					
	1st biweek storage	2nd biweek storage	3rd biweek storage	4th biweek storage	5th biweek storage	6th biweek storage
N (kg ha⁻¹)						
0(Control)	12.13	19.51	29.44	36.84^{bc}	49.31^b	44.77^c
50	11.62	18.93	28.92	35.98^c	50.11^b	46.78^{bc}
100	12.42	19.88	31.27	41.84^{ab}	56.27^a	51.76^{ab}
150	11.51	21.04	33.03	43.10^a	57.10^a	57.46^a
Significance	NS	NS	NS	*	*	***
Genotypes						
Huruta	11.79	19.94	29.34	38.40^b	51.77^b	49.28^b
Negelle	11.82	19.77	28.89	36.36^b	49.34^b	44.86^b
Dz-sht-68	11.95	19.69	29.83	36.89 ^{b}	48.95^b	46.05^b
Local	12.12	19.96	34.60	46.11^a	62.72^a	60.57^a
Significance	NS	NS	NS	**	***	***
SED	0.76	1.87	2.40	2.58	2.82	2.97
CV (%)	15.63	23.06	19.18	16.02	13.00	14.50

Means followed by the same letter(s) within a column are not significantly different at p<0.05; ns= non-significant; ***, **, *= significant at p<0.001, p<0.01 and p<0.05, respectively.

During the last biweek of storage significant interaction effect (p<0.01) between genotypes and N rates were observed on percent weight loss of bulbs (Table 9). At the end of the storage period, application of N at the rate of 150 kg N ha⁻¹ increased the cumulative weight loss in Huruta and local cultivar by about 67%, as compared to the Negelle shallot variety that had the least weight loss percentage. The effect of nitrogen fertilization at 50 and 100 kg N ha⁻¹ did not differ from the unfertilized treatment in all improved shallot genotypes. Similarly, the local cultivar fertilized with 50 kg N ha⁻¹ had low percentage of weight loss which was at par with other genotypes fertilized with N at similar rate.

The storability of shallot under ambient weather conditions appeared to be very low, mainly due to net moisture loss from the tissues under the high ambient temperature. In general, there was high total weight loss in this study due to very high temperature during the storage periods (Appendix Fig. 1). However, in the last two weeks of storage percent weight gain was observed. This could be due to the presence of available moisture in the storage environment or external atmosphere than the internal tissue of the bulb as result of rainfall during that particular time, which might have enhanced conductivity of water vapor into the bulbs, ultimately increasing the bulb weight.

Table 9. Nitrogen rates, genotypes and their interaction effect on cumulative percent weight loss after the three months of storage

Genotypes	N (kg ha^{-1})			
	0	50	100	150
Huruta	38.40^c	44.63^c	46.05^c	68.05^a
Negelle	40.51^c	47.75^c	49.96^{bc}	41.22^c
Dz-sht-68	39.11^c	46.95^c	47.26^c	50.89^{bc}
Local	61.05^{ab}	47.80^c	63.78^a	69.66^a
SE±		4.20		
Significance		**		
CV (%)		14.50		

Means followed by the same letter(s) are not significantly different at $p<0.05$; **= significant at $p<0.01$.
SED(G×N)= 5.94

4.5.2. Rotting of bulbs

Nitrogen fertilization had a significant ($p<0.05$) effect on percentage of bulb rotting on the first, third and fourth biweeks of storage assessment periods (Appendix Table 9 and Table 10). The highest percentage of bulb rottings (2.81%, 5.27% and 6.97%) were observed in a plot fertilized with 150 kg N in the 1st, 3rd and 4th biweeks, respectively (Table 10). Percent rotting in bulbs fertilized with 100 kg N ha^{-1} did not differ significantly from the highest percent loss observed at 150 kg N ha^{-1}. At the end of two months of storage, application of 150 kg N ha^{-1} caused 78.7% higher rotting over the control. Generally, N fertilization tended to increase rotting percentage of bulbs with N rates applied during storage period.

Singh (1987) also reported similar results where rotting of bulbs increased due to an increase in rate of applied nitrogen which could be attributed to the fact that plants supplied with high nitrogen could have encouraged the production of large bulbs with soft succulent tissues which could be susceptible to attacks by rotting micro-organisms. Similar influences of nitrogen on onion were also reported by Vaughan (1960) and Bhalekar *et al.* (1987). On the contrary, Celestino (1961) and Hussien (1996) reported that application of nitrogen did not show any significant effect on rotting of bulbs even if the rotting of bulbs were tended to increase with the increase in the amount of nitrogen applied to the soil.

There were significant differences ($p<0.05$) in percentage of bulb rotting among the shallot genotypes during most part of the storage period (Table 10). A significantly large percent of bulb rotting was recorded for the local cultivar. After 12 weeks of storage, bulb rotting loss ranged from 6.86 to 18.67% in Negelle and local cultivar, respectively. The rotting loss for the local cultivar was 169% higher than that of Huruta up to the end of two and a half months of storage. There was no significant variation in percent bulb rotting among the improved genotypes Huruta, Negelle and Dz-sht-68 throughout the storage period. The highest rotting loss in the local cultivar could be due to thin outer scale of bulbs common to the yellow cultivars which could have favored infection and development of the rotting micro-organisms. The result is in agreement with the finding of Saxena *et al.* (1974) who reported that red skinned cultivars had higher storage potential in comparison with yellow and white skinned onion cultivars.

In the late stage of storage, there was development of black mould due to the sprouting of the fungus. This fungus invades the fleshy scales and the infected scales developed a white fungal mycelium on which sooty, black sporing bodies developed. The optimum temperature for the invasion and growth of this pathogen is about 32.5 °C (Brewster, 1994), which was also experienced during this study (Appendix Table 2).

Table 10. Effect of nitrogen fertilization and genotypes on percent rot loss of marketable bulbs of shallot in ambient storage at Sirinka

Treatment	Cumulative percent rot after					
	1st biweek storage	2nd biweek storage	3rd biweek storage	4th biweek storage	5th biweek storage	6th biweek storage
N (kg ha⁻¹)						
0 (Control)	1.04 (1.18)[c]	2.00 (1.53)	2.35 (1.64)[b]	3.90 (2.03)[b]	6.14 (2.50)	8.03 (2.80)
50	1.29 (1.30)[bc]	2.89 (1.73)	3.48 (1.85)[ab]	4.11 (2.05)[b]	8.41 (2.77)	10.06 (3.01)
100	2.25 (1.58)[ab]	3.57 (1.95)	4.56 (2.17)[a]	6.22 (2.50)[ab]	9.96 (3.09)	11.28 (3.30)
150	2.81 (1.72)[a]	3.87 (2.03)	5.27 (2.31)[a]	6.97 (2.69)[a]	10.56 (3.23)	12.73 (3.55)
Significance	*	NS	*	*	NS	NS
Genotypes						
Huruta	1.45 (1.27)[b]	2.46 (1.62)	3.02 (1.72)[b]	4.05 (2.01)[b]	5.75 (2.38)[b]	7.33 (2.71)[b]
Negelle	1.93 (1.50)[b]	2.57 (1.70)	3.22 (1.85)[b]	4.55 (2.18)[b]	6.18 (2.52)[b]	6.86 (2.64)[b]
Dz-sht-68	1.19 (1.25)[b]	2.83 (1.74)	3.72 (1.97)[ab]	5.08 (2.28)[b]	7.66 (2.76)[b]	9.23 (3.01)[b]
Local	2.82 (1.76)[a]	4.47 (2.18)	5.71 (2.41)[a]	7.52 (2.80)[a]	15.48 (3.93)[a]	18.67 (4.31)[a]
Significance	*	NS	*	*	***	***
SED	0.18	0.20	0.23	0.24	0.29	0.31
CV (%)	30.94	26.93	28.99	25.06	24.31	24.01

Means followed by the same letter(s) within a column are not significantly different at $p<0.05$; ns= non-significant; *, ***= significant at $p<0.05$ and $p<0.001$, respectively.
The data in the bracket are square root transformed values.

4.5.3. Sprouting and rooting of bulbs

In this study the parameters sprouting and rooting did not show difference due to nitrogen fertilization and genotypes as well as interaction effects under ambient condition of storage. Sprouting of bulbs was not recorded throughout the storage period although there were some bulbs that sprouted in the nitrogen fertilized shallot bulbs in comparison with the unfertilized ones. Similarly, sprouting of some bulbs was observed in Dz-sht-68 as compared to the other genotypes. The absence of sprouting of bulbs could be attributed mainly to the curing treatment before storage, the high temperature and low relative humidity in the store that could rather enhance drying of bulbs than sprouting (Appendix Fig. 1). Similar findings were also reported by Hussien (1996) who found that nitrogen fertilizer application did not have significant effect on sprouting and average length of sprout during the four months of storage period. In contrast, Celestino (1961), Bhalekar *et al.* (1987), and Dankhar and Singh (1991) found significant effect on onions interms of sprouting due to N fertilization.

Dankhar and Singh (1991) reported that high dose of nitrogen produced thick-necked bulbs in common onion that increased sprouting in storage due to greater access of oxygen and moisture to the central growing point which was not observed in shallot bulbs in this study. Bhalekar *et al.* (1987) observed that sprouting was increased with increasing nitrogen levels from 0 to 150 kg N ha^{-1}. Celestino (1961) reported that bulbs fertilized with 60 or 120 kg N ha^{-1} sprouted twice as much under common storage as compared to those which were not fertilized.

In general, the shelf life of shallot under ambient weather conditions in this study appeared to be very low, mainly due to high net moisture loss from the tissues and to some extent due to bulb rotting as well as respiratory use of storage substrates. Only about 0.27% of the bulbs sprouted while no rooting loss was observed during the three months of storage (data not shown), which could be due to very low RH and very high temperature.

5. SUMMARY AND CONCLUSION

Shallot is among the most important vegetables grown and utilized in Ethiopia. However, productivity of this crop is substantially low compared to crops like onion. Nutrient supply is known to affect productivity, quality and storability of onions and shallots. In order to improve productivity of shallot, appropriate management practices for different genotypes should be taken into consideration. In addition to increment of yield, improvement of the shelf life has to be achieved during the production stage, coupled with appropriate postharvest handling practices. Hence, nitrogen fertilization rates for released and elite shallot genotypes were considered in this study.

Therefore, the study was conducted to investigate the effect of N fertilization on yield, yield components and shelf life of some shallot genotypes. The treatments consisted of a 4×4 factorial combination of Nitrogen fertilizer (0, 50, 100 and 150 kg N ha^{-1}) and shallot genotypes (Huruta, Negelle, Dz-sht-68 and local cultivar). The experiments were undertaken using randomized complete block design with three replications. The field experiment was carried out at Jari experimental site of Sirinka Agricultural Research Center (SARC) in Wello, Northeastern Ethiopia on a silt-sandy soil. After harvest, 5 kg of marketable bulbs from each treatment were stored for three months at SARC in ambient storage atmosphere of 31.6 °C maximum and 15.8 °C minimum average temperatures and 46% relative humidity during the storage period.

Nitrogen fertilization significantly increased all of the growth parameters considered in this study. Similarly, significant genotypic variations were observed in growth parameters, viz: plant height, number of leaves per plant, leaf diameter, biomass per plant and days to maturity. Nitrogen fertilization had also a significant interaction effect with genotypes on the number of bulb splits per plant, bulb diameter, mean bulb weight, unmarketable bulb yield and percent bulb weight loss during the last biweek of storage.

The interaction effect of local cultivar with N at 100 kg ha^{-1} showed increase in the number of bulb splits by 88.11% and 71.50% compared to Huruta and local cultivar with no fertilizer application, respectively. The application of N at 150 kg ha^{-1} increased the bulb diameter of Dz-sht-68 genotype by 71.74% compared to the local cultivar with no

46

nitrogen application. Similarly, application of 150 kg N ha^{-1} increased the mean bulb weight of Negelle by 175.2% and 108.8% compared to the local cultivar fertilized with 0 and 50 kg N ha^{-1}, respectively.

Among the nitrogen rates applied, the highest and significant values in plant growth and bulb yield and associated traits were observed on shallots fertilized at 100 kg N ha^{-1}. In harmony with these, the relationship between the applied levels of N and yield and yield component parameters of the shallot crop considered in this study were positive and significantly correlated, suggesting that an overall increase in these parameters with increased rate of N application.

Shallot productivity was improved by the application of N fertilizer. Application of N at a rate of 100 kg ha^{-1} significantly increased the marketable and total bulb yields. Genotypic differences in marketable and total bulb yields were very highly significant where Dz-sht-68 produced the highest marketable and total bulb yields while the local cultivar had the lowest yield.

Total bulb yield showed positive and significant statistical correlations with plant height, number of leaves per plant, leaf diameter, days to maturity, biomass per plant, bulb diameter and mean bulb weight. There was also a positive correlation of total bulb yield with number of lateral branches per plant and number of bulb splits per plant, which suggested that genotypes and N fertilization producing large number of lateral branches and leaves could produce large number of bulb splits thereby resulting in increased bulb yield. The result also indicated that shallot genotypes and N fertilization producing relatively large bulb size could increase the total and marketable bulb yields.

The bulb quality parameters, viz: pungency, TSS and dry matter of the shallot bulbs, assessed during the laboratory analysis immediately after harvest, didn't show significant difference due to nitrogen fertilization. However, for TSS and dry matter, genotypes recorded a very highly significant variation where the local genotype had the highest TSS and dry matter contents.

In the early stage of storage, there was a non-significant difference in cumulative percent weight loss of shallot bulbs due to nitrogen fertilization and genotypes. However, in the

47

late stage of storage, cumulative percent weight loss of the bulbs showed a significant ($p < 0.05$) increase due to nitrogen fertilization and also the local cultivar had more percentage weight loss of bulbs. The interaction effects showed that local shallot cultivar fertilized with 100 to 150 kg N ha^{-1} and Huruta fertilized with 150 kg N ha^{-1} had the highest percentage weight loss compared to the remaining treatment combinations.

N fertilization had a significant effect on cumulative percent rot at 1st, 3rd and 4th biweeks where the highest bulb rot recorded in N fertilized plots with 100 and 150 kg ha^{-1}. On the other hand, genotypes showed a significant difference in cumulative percent rot of the bulbs in most of the storage periods except the 2nd biweek where the local cultivar had the highest bulb rot loss.

There was no rooting and sprouting of bulbs recorded during the storage period. However, the overall storage life of shallot bulbs under ambient weather conditions in a grass-thatched roofing house appeared to be very low due to high temperature and low relative humidity that resulted in a very high rate of weight loss of the bulbs.

In conclusion, the results of the study have revealed that increasing the rate of nitrogen significantly increased yield and yield components of shallot but did not influence storability and shelf life. The study has also showed that the four genotypes differed significantly in yield potentials and storability.

However, to obtain concrete and reasonable results on effects of nitrogen fertilization on yield, quality and storability of shallot, it would be valuable to repeat the study under different environments and seasons to arrive at a sound conclusion. Also, evaluation of shelf life of different shallot genotypes produced under different management practices deserve further investigation under alterative storage methods in order to get solution for the naturally very low storability of the bulbs under ambient storage conditions.

6. REFERENCES

Abayneh, M., 2001. Variability and association among bulb yield, quality and related traits in onion (*Allium cepa* L.). M.Sc. Thesis Presented to the School of Graduate Studies of Alemaya University. 51p.

Abbey, L., 2004. Electronic nose evaluation of onion headspace volatiles and bulb quality as affected by nitrogen, sulphur and soil type. *Journal of Vegetable Crop Production,* 145(1): 41-50.

Abbey, L., R. Kanton and H. Braimah, 1998. Susceptibility of shallot to the timing and severity of leaf damage. *Journal of Horticultural Science and Biotechnology, 73: 803-805.*

Ali, A.A. and A.M. Shabrawy, 1979. Effect of some cultural practices and some chemicals on the control of neck rot disease caused by *Botrytis allii* during storage and in the field for seed onion production in A.R.E., *Agric. Res. Rev. Plant pathol.* 57: 103.

Allard, R.W. 1960. Principles of Plant Breeding. John Willey and Sons Inc., New York. 485p.

Archer, J., 1988. Crop Nutrition and Fertilizer Use. Second Edition. Farming Press Ltd. Wharfedaale Road, Ipswich, Suffolle.

Arifin, S. and H. Okubo, 1996. Geographical distribution of allozymes patterns in shallot (*Allium cepa.* var. *ascalonicum* Backer) and wakegi onion *A.cepa* x (wakegiAraki). *Euphytica,* 91: 305-313.

Barta, S.K., G. Kalloo and B. Singh, 1983. Combining ability, heterosis and analysis of phenotypic variation in onion. *Haryana J. Hort. Sci.* 12: 119.

Batal, K.M., K. Bondari, D.M. Ganberry and B.G. Mullinix, 1994. Effect of source, rate and frequence of N application on yield, marketable grades and rot incidence of sweet onion (*Allium cepa* L. cv. Granex-33). *J. of Hort. Sci.* 69: 1043-1051.

Bayu, W., A. Getachew, and T. Mamo, 2002. Response of sorghum to nitrogen and phosphorus fertilization in semi-arid environments in Wello, Ethiopia. *Acta Agron Hungurica,* 50: 53-65.

Bendarz, F., K. Kadam and G.O. Dosunmu, 1986. Yield and quality of various short day onion varieties (*Allium cepa* L.) in northern Nigeria in relation to date of sowing, in Nortson 8th Annual conference/10th Anniversary of Nihort, Ibadan.

Bhalekar, M.N., P.B. Kale and L.V. Kulwal, 1987. Storage behavior of some onion varieties (*Allium cepa* L.) as influenced by nitrogen levels and preharvest spray of maleic hydrazide. *Journal Pkv Res.* 11(1): 38-46.

Bielinska-Czarnecka M., A. Kepkowa, E. Kielak and E. Zdanowska, 1981. Influence of size of onion bulb cv. "Czerniakowska" on its dormancy, sprouting and rooting, *Acta Agrobotanica, Warsaw,* 34: 129.

Black, C.A., 1965. Determination of exchangeable Ca, Mg, Na, K, Mn and effective cation exchange capacity in soil. Methods of soil analysis. 9(2) *American Society of Agronomy*, Madison, Wisconsin.

Bremner, J.M. and C.S. Mulvancy, 1982. Nitrogen total, pp. 595-624. In: A.L. Page (ed.). Methods of soil analysis, part two, Chemical and microbiological properties. 2nd ed. *American Society of Agronomy*, Maidison, Wisconsin.

Brewster, J.L., 1987. The effect of temperature on the rate of sprout growth and development within stored onion bulbs. *Annals of Applied Biology,* 111: 463-465.

Brewster, J.L., 1990. Physiology of crop growth and bulbing. pp. 53-88. In: H.D. Rabinowitch and J.L. Brewster (eds.). Onions and Allied Crops, Botany, Physiology and Genetics, Vol. I. CRC Press, Boca Raton, Florida.

Brewster, J.L., 1994. Onions and Other Vegetable Alliums. CABI Publishing. Wallingford, UK. 236p.

Brewster, J.L., H.R. Rowse and A.D. Bosch, 1991. The effects of sub-seed placement of liquid N and P fertilizer on the growth and development of bulb onions over a range of plant densities using primed and non-primed seed. *J. Hort. Sci.* 66: 551-557.

CACC, 2002. (Central Agricultural Census Commission Part I): Report on the preliminary results of area, production and yield of temporary crops. Addis Ababa.

Carter, M.R., 1993. Soil sampling and methods of analysis. *Canadian Soil Science.*

Celestino, A.F., 1961. The effect of irrigation, nitrogen fertilization and maleic hydrazide on the yield, composition and storage behavior of bulbs of two onion varieties. *Journal of Philippines Agriculture,* 44: 479-501.

Cizauskas, A., P. Viskelis, R. Dris, O.I. Oladele, 2003. Influence of nitrogen rates on onion yield, quality and storability. *Moor Journal of Agricultural Research*, 4(1): 85-89.

Currah, L. and F.J. Proctor, 1990. Onion in Tropical Regions. *Natural Resources Institute*, UK. Bul. No. 35.

Currah, L., 2002. Onions in the tropics: cultivars and country reports. pp. 379-407. *In*: H.D. Rabinowitch and L. Currah (eds.). Allium Crop Science: Recent Advances. CABI Publishing, London.

Dankhar, B.S. and J. Singh, 1991. Effect of Nitrogen, Potash and Zinc on storage loss onion bulbs (*Allium cepa* L.). *Journal of Vegetable Science*, 18: 16-23.

Darbyshire, B. and R.J. Henry, 1981. The distribution of fructans in onions. *Journal of New phytol. 81: 29-34.*

Day, P.R., 1965. Hydrometer method of particle size analysis. pp. 562-563. In: Black C.A. (ed.). Methods of Soil Analysis. Agronomy. Part II, No. 9. *American Society of Agronomy*, Madison, USA.

De Visser, C.I.M., 1998. Effect of split application of nitrogen on yield and nitrogen recovery of spring-sown onions and on residual nitrogen. *Journal of Horticultural Science and Biotechnology,* 73: 403-411.

Dewis, J. and P. Freitans, 1970. Physical and chemical methods of soil and water analysis. Bull. No. 10. FAO, Rome, Italy.

Epstein, E. and R.K. Jefferies, 1964. The Genetic Bases of Selective Ion Transport in Plants. *Ann. Rev. Pl. Physiol.* 15: 169-84.

Fasika, S., 2004. Variability and association among bulb yield, yield components and quality parameters in shallot (*Allium cepa* var *aggregatum* Don.) genotypes of Ethiopia. M.Sc. Thesis Presented to School of Graduate Studies of Alemaya University. 83p.

Fritsch, R.M. and N. Friesen, 2002. Evolution, domestication and taxonomy in Alliums. pp5-30. *In*: H.D. Rabinowitch and L. Currah (eds.). Allium Crop Science: Recent advances. CABI Publishing, London.

Getachew, T. and Z. Asfaw, 2000. Research achievements in garlic and shallot. Research Bull. Report No. 36. EARO, Ethiopia.

Getachew, T., 1996. The effect of vernalization on bolting and flowering of shallot. M.Sc. thesis submitted to College of Agriculture, Alemaya University.

Getahun, D., A. Zelleke, E. Derso and E. Kiflu, 2007. Storability of shallot cultivars (*Allium cepa* L. var. *ascalonicum* Baker) at Debre Zeit. Acta Horticulturae.

Gomez, A.K. and A.A. Gomez, 1984. Statistical procedure for agricultural research 2nd (Ed.). A Wiley Inter-Science Publication, New York.

Gorin, N. and S. Borcsok, 1980. Chemical composition of stored onions, cultivar Hydro, as a criterion of freshness. *Lebensm-Wiss. Technol.* 20: 461-463.

Green, J. H., 1972. Suggestions for improved storage of onions in Northern Nigeria, Samaru Agric. Newsl. 14: 56.

Grubben, G.J.H., 1994. Constraints for shallot, garlic, and Welsh onion in Indonesia: a case study on the evolution of Allium crops in the equatorial tropics. *Acta Horticulturae,* 358: 333-339.

Gubb, I.R. and M.S.H. Tavish, 2002. Onion preharvest and postharvest considerations. pp. 237-250. In: H.D. Rabinowitch and L. Currah (eds.). Allium crop science. CABI publishing, UK.

Hassan, M.S. and A.T. Ayoub, 1978. Effect of nitrogen, phosphorus and potassium on yield of onion in the Sudan. Gezira. *Experimental agriculture,* 6: 345-350.

Havey, M.J., 1993. Onion (*Allium cepa* L.). pp. 35-49. *In:* G. Kallo and B.O. Bergh (eds.). Genetic Improvement of Vegetable Crops. Pergmon Press, Oxford.

Hay, R.K.M. and A.J. Walker, 1989. An introduction to the physiology of crop yield. John Wiley & Sons, Inc., New York.

Hayashi, Y. and R. Hattano, 1999. Annual nitrogen leaching to subsurface water from clayey aquic soil cultivated with onions in Hokkaido, Japan. *Soil Science and Plant Nutrition*, 45: 451-459.

Hayden, N.J. and R.B. Maude, 1997. The use of integrated pre- and post-harvest strategies for the control of fungal pathogens of stored temperate onions. *Acta Horticulturae, 433: 475-479.*

Hegde, D.M., 1988. Effect of irrigation and nitrogen fertilization on yield, quality, nutrient uptake and water use of onion (*Allium cepa L.*). *Singapore Journal of Primary Industries,* 16: 111-123.

Hegde, D.M., 1986. Effects of irrigation and nitrogen fertilization on water relations, canopy temperature, yield nitrogen uptake and water use of onions. *Indian Journal of Agricultural Sciences,* 56: 858-867.

Henriksen, K., 1987. Effect of N- and P-fertilization on yield and harvest time in bulb onions (*Allium cepa L.*). *Acta Horticulturae,* 208: 207-215.

Henry, L.T. and J.C.D. Raper, 1989. Effects of Root Zone Acidity on Utilization of Nitrate and Ammonium in Tobacco Plants. *J. Pl. Nutr.* 12(7): 811-826.

Hesse, P.R., 1972. A textbook of soil chemical analysis. John Murry Limited, London, Great Britain. 470p.

Hurst, W.C., R.L. Shewfelt and G.A. Schuler, 1985. Shelf life and quality of changes in summer storage of onions (*Allium cepa*). *J. Food Sci.* 50: 761.

Hussaini, M.A., Amans, E.B. and A.A. Ramalan, 2000. Yield, bulb size distribution, and storability of onion (Allium cepa L.) under different levels of N fertilization and irrigation regime. *Tropical agriculture*, 77: 145-149.

Hussien, J., 1996. Influence of nitrogen and maleic hydrazide on the keeping quality and subsequent establishment of shallot (*Allium cepa* L. var. *ascalonicum*). M.Sc. Thesis submitted to Alemaya University of Agriculture. 64p.

Isenberg, F.M.R., T.H. Thomas, A.M. Pendegrass and M. Abdel-Rahman, 1974. Hormone and histological differences between normal and maleic hydrazide treated onions stored over winter. *Acta Horticulturae, 38: 95-125.*

John, M.S., 1992. Producing Vegetable Crops. Interstate Publisher Inc. 403p.

Jones, H.A., and L.K. Mann, 1963. Onions and their allies: Botany, cultivation, and utilization. New York. 285p.

Kafkafi, U.W. and S. Genbaum, 1971. Effect of Potassium Nitrate on Growth, Cation Uptake and Water Requirement of Tomato Grown in Sand Culture. *Israel J. Agri. Res.* 21: 13-20.

Kato, T., 1966. Physiological studies on the bulbing and dormancy on onion plants. V. Relations between dormancy and organic constituents of bulbs. *J. Jpn. Soc. Hort. Sci.* 35: 143.

Kato, T., M. Yamagata and S. Tsukahra, 1987. Nitrogen nutrition, its diagnosis and postharvest bulb rot in onion plant. *Bulletin of the Shikoku National agricultural Experiment station, 48: 26-49.*

Kebede W., U. Gretsson, and J. Ascard, 2003a. Shallot yield, quality and storability as affected by irrigation and nitrogen. *Journal of horticultural science and biotechnology,* 78(4): 549-553.

Kebede, W., U. Gretsson, and J. Ascard, 2003b. Response of shallots to mulching and nitrogen fertilization. *Hort. Sci.* 38(2): 1-5.

Kebede, W., U. Gretsson, and J. Ascard, 2002a. Response of shallots to N, P, and K fertilizer rates. *Trop. Agri. (Trinidad), 79*(4): 205-210.

Kebede, W., U. Gretsson, and J. Ascard, 2002b. Season, and nitrogen source and rate affect development and yield of shallot. *Journal of vegetable crop production,* 8: 71-81.

Ketter, C. A. T. and W.M. Randle, 1998. Pungency assessment in onions. pp. 177-196, Volume 19. Proceedings of the 19th Workshop/Conference of the association for biology laboratory education (ABLE), 365 pages, Univesity of Georgia, Athens, Georgia.

Kleinkopf, G.E., D.T. Westermann, M.J. Wille and G.D. Kleinscmidt, 1987. Specific Gravity of Russet Burbank Potatoes. *American Potato Journal.* 64: 579-587.

Komochi, S., 1990. Bulb Dormancy and Storage pp. 89-91. In: H.D. Tabinowitch and J.L. Brewster (eds.). Onions and Allied Crops. Vol. II. Botany, Physiology and Genetics. CRC Press, Boca Raton, Florida.

Konesky, D.W., M.Y. Siddiqi, A.D.M. Glass and A.I. Hsiao, 1989. Genetic Differences Among Barley Cultivars and Wild Oats Lines in Endogenous Seed Nutrient Levels; Initial Nitrate Uptake Parts and Growth in Relation to Nitrate Supply. *J. Pl. Nutr.* 12(1): 9-35.

Lancaster, J.E. and M.J. Boland, 1990. Flavor Biochemistry. Vol. III. pp. 33-72. *In:* J.L. Brewster and H.D. Rabinowitch (Eds.). Onion and Allied Crops., Inc., Boca Raton. CRC Press. Florida.

Le Gouis, J., D. Beghin, E. Heumez and P. Pluchard, 2000. Genetic differences for nitrogen uptake and nitrogen utilization efficiencies in winter wheat. *Europ. J. Agron.* 12: 163-173.

Lemma Desalegn and Shimels Aklilu, 2003. Research Experiences in Onion Production. EARO. Addis Ababa.

Lin, M., J.F. Watson and J.R. Baggett, 1995. Inheritance of soluble solids and pyruvic acid content of bulb onions. *J. Amer. Soc. Hort. Sci.* 120: 119-122.

Lutz, J.M. and R.E. Hardenburg, 1968. The commercial storage of fruit, vegetables, florist and nursery stock. USDA Agriculture Handbook, Vol. 66. U.S. Government Printing Office, Washington D.C.

Maier, N.A., A.P.Dahlenburg and T.K. Twigden, 1990. Effect of nitrogen on the yield and quality of irrigated onions (*Allium cepa* L.) cv. Cream Gold grown on silicaceous sands. *Australian Journal of Experimental Agriculture,* 30: 845-851.

Maini, S.B., B. Diwan and J.C. Anand, 1984. Storage behavior and drying characteristics of commercial cultivars of onion, *J. Food Sci. Technol.* (India), 21: 417.

Mamo, T., I. Haque and C.S. Kamara, 1988. Phosphorus status of some Ethiopian highland vertisols in Sub-Saharan Africa. In Proc. conf. held at ILCA, 31 Aug. - 4 Sept., 1987, Addis Ababa, Ethiopia.

Marschner, H., 1995. Mineral Nutrition of Higher Plants, 2nd ed. Academic press. London. 196p.

Maude, R.B., N.F. Lyons, L. Gurd, Abu Baker, H. EI-Muallem and D. Bamakrama, 1991. Disease problems of onions in the Republic of Yemen. *Onion Newsletter for the tropics, 3: 34-38.*

Mc Cullough, D.E., P.H. Girardin, M. Mihajlovic, A. Aguilera and M. Tollenaar, 1994. Influence of N supply on development and dry matter accumulation of an old new maize hybrid. *Can. J. Plant Sci.* 74: 471-477.

Metasebia Merid and Shimels Hussein, 1998. Proceeding of the 15th Annual research and extension review meeting, 2 April 1998. Alemaya Research Centre. Alemaya University of Agri. pp. 216-235.

Miller, R.W. and R.L. Donanue, 1995. Soils in Our Environment. (7th ed.). Prentice Hall, Englewood Cliff.

Mondal, M.F. and M.H.R. Pramanik, 1992. Major factors affecting the storage life of onion. *International Journal of Tropical Agriculture, 10: 140-146.*

Muchow, R.C. and R. Davis, 1988. Effect of nitrogen supply on the comparative productivity of maize and sorghum in a semi-arid tropical environment: II. Radiation interception and biomass accumulation. *Field Crops Res.* 18: 17-30.

Muchow, R.C., 1988. Effect of nitrogen supply on the comparative productivity of maize and sorghum in semi-arid tropical environment. I. Leaf growth and leaf nitrogen. *Field Crops Res.* 18: 1-16.

Muchow, R.C., 1994. Effect of nitrogen on yield determination in irrigated maize in tropical and subtropical environments. *Field Crops Res.* 38: 1-13.

Murphy, H.F., 1959. A Report on Fertility Status of Some Soils of Ethiopia. HSIU/College of Agric. Alemaya. Experiment Station. Bull. No. 1.

Murphy, H.F., 1968. A Report on Fertility Status of Some Soils of Ethiopia. HSIU/College of Agric. Alemaya. Experiment Station. Bull. No. 44.

Nasreen, S., M.M. Haque, M.A. Hossain and A.T.M. Farid, 2007. Nutrient uptake and yield of onion as influenced by nitrogen and sulphur fertilization. *Bangladesh Journal of Agricultural Research,* 32(3): 413-420.

Neeteson, J.J., R. Booij and A.P. Whitmore, 1999. A review of sustainable nitrogen management in intensive vegetable production systems. *Acta Horticulturae,* 506: 17-28.

Novoa, R. and R.S. Loomis, 1981. Nitrogen and plant production. Plant Soil. 58: 177-204.

Olsen, S.R., C.V. Cole, F.S. Watanabe and L.A. Dean, 1954. Estimation of available phosphorus in soils by extraction with sodium bicarbonate. USA Circular, 939: 1-19.

Pathak, C.S., 1994. Allium improvement for the tropics: problems AVRDC strategy. *Acta Horticulturae, 358: 395-400.*

Peiris, K.H.S., J.L. Mallon and S.J. Kays, 1997. Respiratory rate and vital heat of some specialty vegetables at various storage temperatures. *Horticulture Technology, 7: 46-49.*

Pire, R., H. Ramire, J. Riera and T.N. De Gomez, 2001. Removal of N, P, K, and Ca by an onion crop (*Allium cepa* L.) in silty-clay soil, in a semiarid region of Venezuela. *Acta Horticulturae, 555: 103-109.*

Proctor, F.J., J.P. Goodlife and D.G. Coursey, 1981. Post-harvest losses of vegetables and their control in tropics, in Vegetable Productivity, Spedding, C.R.W., Ed., Macmillan, London, 139p.

Rao, M.R., A. Niag, F. Kwcsiga, B. Duguma, S. Frazel, B. Jama and R. Buresh, 1998. Soil Fertility Replenishment in Sub-Saharan Africa. pp. 2-5. D. Londoen (ed.). In Agroforestry Today. Vol. 10. No. 2. International Center for Research in Agroforestry (ICRAF), UK.

Reiley, H.E. and C.L. Jr. Shry, 1979. Introductory Horticulture. Delmar Publisher. Albany, New York.

Rice, R.P., L.W. Rice and H.D. Tindall, 1993. Fruit and vegetable production in warm climates. The Macmillan press Ltd. London and Basingstoke.

Richardson, H.L., 1968. The Use of Fertilizers: The Soil Resources of Tropical Africa. R.P. Moss (ed.). Cambridge University Press. 138p.

Rowell, D.L., 1994. Soil science: Method and applications. Addison Wesley Longman Limited, England. 350p.

Rubatzky, V.E. and M. Yamagunchi, 1997. World Vegetables; Principles, Production and Nutritive Value 2nd ed. International Thomson publishing. 804p.

Rutherford, P.P. and R. Whittle, 1982. The carbohydrate composition of onions during longterm cold storage. *J. of Hort. Sci. 57: 349-356.*

Rutherford, P.P. and R. Whittle, 1984. Methods of predicting the long-term storage of onions. *J. of Hort. Sci. 59: 537-543.*

Saimbhi, M.S. and K.S. Randhawa, 1982. Losses in white onions variety Punjab-48 under ordinary storage conditions as influenced by bulb size, *J. Res. India*, 19: 188.

Salunkhe, D.K. and S.S. Kadam, 1998. Handbook of Vegetable Science and Technology: Production. Technology and Engineering. 742p.

Sanchez, P.A., 1976. Properties and Management of Soils in the Tropics. John Wiley and Sons, New York.

SARC (Sirinka Agricultural Research Center), 2007. Annual Progress Report. Sirinka, Northeastern Ethiopia.

Saxena, G.K., L.H. Halsey, D.D. Gull and N. Persuad, 1974. Evaluation of carrot and onion cultivars for commercial production in Guyana. *Hort.* Sci. 2: 257.

Schwimmer, S. and W.J. Weston, 1961. Enzymatic development of pyruvic acid in onion as a measure of pungency. *J. Agric. Food Chem.* 9: 301-304.

Sebsebe, Z., 2006. Effect of nitrogen levels, harvesting time and curing on bulb quality and shelf life shallot (*Allium cepa* var. *aggregatum* Don.). M.Sc. Thesis Presented to the School of Graduate Studies of Alemaya University. 77p.

Seifu, G., 1981. Research in horticultural crops in Ethiopia from 1977-1981. IAR, Addis Ababa.

Shimels, A., 1998. Prospects of growing shallots from true seed under irrigation in the Rift Valley. *AgriTopia,* 13: 8-9.

Singh, J., 1987. Storage studies in onion. Ph.D. Thesis, Dept of Veg. Crops CCS; Haryana Agricultural University, Hisar, India.

Singh, R.P., 1981. Genetic evaluation and path analysis in onion. *Madras Agric. J.* 68: 61-68.

Sørensen, J.N. and K. Grevsen, 2001. Sprouting in bulb onions (*Allium cepa* L.) as influenced by nitrogen and water stress. *Journal of Horticultural Science and Biotechnology,* 76: 501-506.

Sørensen, J.N., 1996. Improve in N efficiency in vegetable production by fertilizer placement and irrigation. *Acta Horticulturae, 428: 131-140.*

Steer, B.T., 1980. The bulbing response to day length and temperature of some Australian cultivars of onion (*Allium cepa* L.). *Australian Journal of Agricultural Research,* 31: 511-518.

Stevenson, R.C. and J.A. Cutcliffe, 1982. Onion variety trials, Charlotte town, 1978-1980. *Agriculture Canada.* Canadex. 258: 34.

Stow, J.R., 1976. The effect of defoliation on storage potential of bulbs of the onion (*Allium cepa* L.). *Annals of Applied Biology, 84: 71-79.*

Sulafa, K.M., A.H. Habish, A.A. Abdalla and A.B. Adlan, 1973. Problems of onion in the sudan, *Trop. Sci.* 15: 319.

Sumiati, E., 1994. Response of shallot and garlic to different altitudes. *Acta Horticulturae, 358: 395-400.*

Suojala, T., T. Salo and R. Pessala, 1998. Effect of fertilization and irrigation practices on yield, maturity, and storability of onions. *Agricultural and Food Science in Finland, 7: 477-489.*

Suzuki, M. and J.A. Cutcliffe, 1989. Fructans in onion bulbs in relation to storage life. *Can. J. Plant Sci.* 69: 1327-1333.

Take, T. and H. Otsuka, 1967. Flavor compounds in various foods. XI. The tasty substances in onion. *Chem. Abstr.* 68(4): 2.

Thompson, A.K., R.H. Booth and F.J. Proctor, 1972. Onion storage in tropics, *Trop. Sci.* 14: 19.

Thompson, H.C. and W.C. Kelly, 1959. Vegetable Crops. McGraw-Hill Book Company, Inc. London. 370p.

Tindall, H.D., 1983. Vegetables in the tropics. Macmillan ed. Ltd. Hampshire and London. 523p.

Tisdale, S.L., W.L. Nelson, J.D. Beaton and J.L. Halvin, 1995. Soil Fertility and fertilizers (5th ed.). Macmillan Publishing Co., Inc. New York.

Uzo, J.O. and L. Currah, 1990. Cultural systems and agronomic practices in tropical climates. pp. 49-62. In: Ranbinowitch, H.D. and J.L. Brewster (Eds.), Onion and Allied crops, Vol. 2. Chapter 3. Agronomy, biotic interactions, pathology and crop protection. CRC Press, Boca Raton. Florida.

Vaughan, E.K., 1960. Influence of growing, curing and storage practices on development of neck rot in onions, Phytopathology, 50: 87.

Walkley, A. and C.A. Black, 1934. Determination of organic matter in the soil by chromic acid digestion. Soil Sci. 63: 251-264.

Waller, M.M. and J.N. Corgan, 1992. Relationship between pyruvate analysis and flavour perception for onion pungency determination. *Hort. Sci.* 9: 301-304.

Ward, C.M., 1979. The effect of bulb size on the storage shelf life of onions. *Ann. Appl. Biol.* 91: 113.

Waskar, D.P., R.M. Khedlar and V.K. Garande, 1999. Effect of postharvest treatment on shelf life and quality of pomegranate in evaporative cool chamber and ambient conditions. *Journal of Food Science and Technology, 2(36): 114-117.*

Wayse, S.B., 1967. Effect of N, P and K on yield and keeping quality of onion bulbs. *India J. Argon.* 12: 379-382.

Welsh, J.R., 1981. Fundamentals of Plant Genetics and Breeding. Willey, New York, 299p.

Wiedenfeld, R., 1994. Nitrogen rate and timing effects on onion growth and nutrient uptake in a subtropical climate. *Subtropical Plant Science*, 46:32-37.

Yamaguchi, M., K.H. Pratt and L.L. Morris, 1957. Effect of storage temperature on keeping quality and composition of onion bulbs and on subsequent darkening of dehydrated flakes, *Proc. Am. Soc. Hort. Sci.* 69: 421.

Yohannes Abebe, 1987. Current activities research recommendation and future strategies of onion research in Ethiopia. pp. 358-369. In: Proceeding of 19th national crop improvement conference. 22-26 April 1987. Institute of Agricultural Research. Addis Ababa, Ethiopia.

Zafrir, G., 1992. Effect of amount and distribution of nitrogen fertilization on the yield, quality and keeping ability of the bulb onion (*Allum cepa* L.). Ph.D. Thesis, The Hebrew University of Jerusalem, Israel.

Zaharah, A., P. Vimala, R. Siti Zainab and H. Salbiah, 1994. Response of onion and shallot to organic fertilizer on bris (rudua series) soil in Malaysia. *Acta Horticulturae*, 358: 429-433.

7. APPENDICES

Appendix I. Soil Analysis

Appendix Table 1. Physico-chemical characteristics of the experimental soil

Soil property	Quantity/type
Particle size distribution (%)	
Clay	26.25
Silt	42.5
Sand	31.25
Textural class	Silt-sandy
pH	7.12
Organic matter (%)	3.01
Available phosphorus (ppm)	8.73
Total nitrogen (%)	0.165
Exchangeable base (meq (+) 100 gm^{-1} soil)	
Ca	28.22
K	2.25
Na	4.17
Cation exchange capacity (meq (+) 100 gm^{-1} soil)	32.04

Appendix II. Meteorological Data for Sirinka

Appendix Table 2. Temperature, relative humidity and rain fall data of Sirinka station for the months Sep. 2007 to July 2008

		Sep.	Oct.	Nov.	Dec.	Jan.	Feb.	Mar.	Apr.	May	Jun	July
T (^{0}C)	Max	26.9	25.6	24.5	24.4	24.1	24.7	28	28.3	30.1	29.9	29.6
	Min	14.3	11.3	10.9	8.9	11.6	10.8	11.3	14.1	16.2	16.5	16.4
RF (mm)		86.9	48.2	0	0	42.8	0	0	27.1	21.8	56.9	-
RH		71	71.3	-	63.4	64.9	55.7	33.5	44.6	44.6	47.4	57.4

Appendix Figure 1. Temperature and relative humidity data of the storage house for the months of April to July 2008

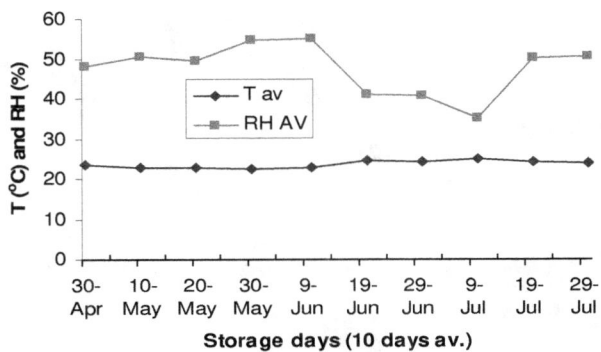

Appendix III. Bulb Size Distribution and Storage Parameter

Appendix Table 3. Effect of nitrogen rate and genotype on bulb size distribution of shallot

Treatment	Bulb size distribution (%)		
	Small	Medium	Large
N (kg ha^{-1})			
0 (Control)	29.58	61.21	9.20
50	32.49	55.90	11.61
100	30.33	59.73	9.94
150	29.64	58.47	11.90
Genotypes			
Huruta	28.68	59.00	12.31
Negelle	23.43	66.44	10.13
Dz-sht-68	24.19	61.52	14.28
Local	65.04	34.61	0.35

Appendix Figure 2. Cumulative percent weight loss of stored shallot bulbs at ambient storage condition (Average 31.6 °C max. and 15.8 °C min. T. and 46% RH) at Sirinka

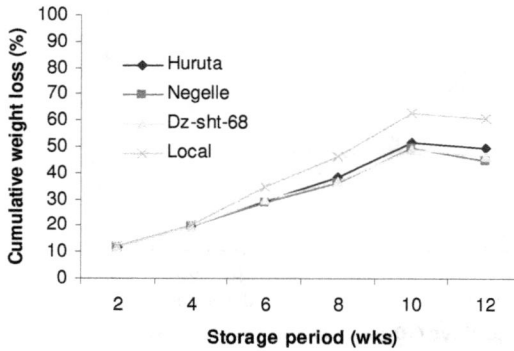

Appendix IV. Analysis of Variance and Correlation Tables

Appendix Table 4. Mean squares for plant height, number of leaves per plant, leaf diameter, number of lateral shoots per plant, days to maturity and biomass per plant of the shallot crop

Source of variations	Degrees of freedom	Mean squares					
		Plant height	No. of leaves	Leaf diameter	Lateral shoots	Days to maturity	Biomass
Block	2	14.45	1266.44***	1.06	9.38***	4.75	0.07***
N	3	297.72***	444.50***	3.63***	1.70**	192.5***	0.11***
G	3	224.1***	1078.22***	3.60***	0.88	649.28***	0.14***
N*G	9	11.61	62.36	0.28	0.21	6.07	0.004
Error	30	14.01	58.75	0.47	0.32	2.75	0.004
Total	47						

** and ***= significant at $p < 0.01$ and $p < 0.001$, respectively

Appendix Table 5. Mean squares for number of bulb splits, bulb diameter and mean bulb weight of the shallot bulb

Source of variations	Degrees of freedom	Mean squares		
		Bulb splits	Bulb diameter	Bulb weight
Block	2	4.002**	124.61***	116.35***
N	3	24.29***	65.64**	127.35***
G	3	14.77***	396.38***	424.24***
N*G	9	3.11***	27.72*	12.45*
Error	30	0.47	9.23	5.15
Total	47			

*, ** and ***= significant at p<0.05, p<0.01and p<0.001, respectively

Appendix Table 6. Mean squares for marketable, unmarketable and total bulb yields and harvest index of the shallot bulb

Source of variations	Degrees of freedom	Mean squares			
		Marketable yield	Unmarketable yield	Total yield	Harvest index
Block	2	25.749	0.037	24.731	64.79
N	3	68.173**	0.246**	75.266**	130.34
G	3	820.128***	1.734***	750.829***	39.50
N*G	9	3.042	0.268***	3.418	30.69
Error	30	11.839	0.043	11.605	61.95
Total	47				

** and ***= significant at p<0.01and p<0.001, respectively

Appendix Table 7. Mean squares for bulb dry matter, pyruvate content and total soluble solids of the shallot

Source of variations	Degrees of freedom	Mean squares		
		Dry matter	Pyruvate content	TSS
Block	2	0.94	0.11	0.23
N	3	1.45	0.41	2.05
G	3	22.71***	2.76	5.46***
N*G	9	1.58	0.13	1.51
Error	30	1.05	1.05	0.70
Total	47			

***= significant at p<0.001

Appendix Table 8. Mean squares of cumulative percentage of weight loss of marketable bulbs of shallot at ambient storage

Source of variations	Degrees of freedom	Mean squares of percent weight loss					
		1st biwk storage	2nd biwk storage	3rd biwk storage	4th biwk storage	5th biwk storage	6th biwk storage
Block	2	5.75	22.84	5.12	5.06	3.54	34.74
N	3	2.21	9.49	42.03	151.43*	197.24*	385.25***
G	3	0.27	0.20	84.43	246.04**	502.97***	616.48***
N*G	9	8.02	16.61	26.09	41.71	60.10	171.44**
Error	30	3.47	20.93	34.58	39.90	47.80	52.95
Total	47						

*, ** and ***= significant at p<0.05, p<0.01 and p<0.001, respectively

Appendix Table 9. Mean squares of cumulative percentage of rotting loss of marketable bulbs of shallot at ambient storage

Source of variations	Degrees of freedom	Mean squares of percent rot loss					
		1st biwk. storage	2nd biwk. storage	3rd biwk. storage	4th biwk. storage	5th biwk. storage	6th biwk. storage
Block	2	0.08	0.05	0.14	0.35	0.76	0.42
N	3	0.74*	0.62	1.11*	1.31*	1.30	1.31
G	3	0.68*	0.75*	1.08*	1.37*	5.94***	7.34***
N*G	9	0.25	0.38	0.58	0.21	0.40	0.45
Error	30	0.20	0.24	0.33	0.34	0.50	0.58
Total	47						

* and ***= significant at p<0.05 and p<0.001, respectively

Appendix Table 10. Simple correlation coefficients of growth parameters, bulb characters and yields of shallot

Traits	PH	NLPP	LD	NLSPP	DTM	TBPP	NBSPP	BD	MBW	PUN	TSS	BDM	HIPP	MBYH	UBYH	TBYH
PH	1															
NLPP	0.59**	1														
LD	0.70***	0.69***	1													
NLSPP	0.06	0.44**	0.27	1												
DTM	0.76***	0.68***	0.66***	-0.003	1											
TBPP	0.70***	0.78***	0.74***	0.33*	0.84***	1										
NBSPP	0.35*	0.24	0.28	0.45**	0.05	0.36*	1									
BD	0.55***	0.73***	0.58***	0.27	0.72***	0.74***	0.22	1								
MBW	0.67**	0.77**	0.71***	0.20	0.82***	0.88***	0.30*	0.81***	1							
PUN	-0.26	-0.35*	-0.17***	0.014	-0.41**	-0.39***	0.04	-0.34*	-0.39**	1						
TSS	-0.07	-0.09	-0.002	0.28	-0.33*	-0.18	0.31*	-0.26	-0.29*	0.26	1					
BDM	-0.56***	-0.47***	-0.49***	0.21	-0.68***	-0.51***	0.09	-0.48***	-0.60***	0.31*	0.26	1				
HIPP	-0.39***	-0.36*	-0.43**	-0.20	-0.33*	-0.36*	-0.26	-0.29***	-0.29*	-0.03	-0.22	0.24	1			
MBYH	0.70***	0.69***	0.67**	0.05	0.89***	0.79***	0.05	0.71***	0.82***	-0.39**	-0.41**	-0.61***	-0.61***	1		
UBYH	-0.18	-0.38**	-0.30*	0.14	-0.42**	-0.31*	0.27	-0.48***	-0.33*	0.12	0.43**	0.36*	0.22	-0.58***	1	
TBYH	0.71***	0.69***	0.67***	0.06	0.90***	0.80***	0.07	0.71***	0.83***	-0.40**	-0.40**	-0.61***	-0.25	0.99***	-0.54***	1

*, ** and *** Correlation is significant at p < 0.05, p < 0.01 and p < 0.001 level, respectively (2 tailed).

PH= Plant height, NLPP= Number of leaf per plant, LD= Leaf diameter, NLSPP=Number of lateral shoots per plant, DTM= Days to maturity, TBPP= Total biomass per plant, NBSPP= Number of bulb splits per plant, BD= Bulb diameter, MBW= Mean bulb weight, PUN= Pungency, TSS= Total soluble solids, BDM= Bulb dry matter, HIPP= Harvest index per plant, MBYH= Marketable bulb yield per hectare, UBYH= Unmarketable bulb yield per hectare, TBYH= Total bulb yield per hectare